Lecture Notes on Mechanics
Intermediate Level

Lecture Notes
on Mechanics
Intermediate Level

LOCK YUE CHEW
Nanyang Technological University, Singapore

ELBERT CHIA
Nanyang Technological University, Singapore

 **World Scientific**

V JERSEY · LONDON · SINGAPORE · BEIJING · SHANGHAI · HONG KONG · TAIPEI · CHENNAI · TOKYO

Published by

World Scientific Publishing Co. Pte. Ltd.

5 Toh Tuck Link, Singapore 596224

USA office: 27 Warren Street, Suite 401-402, Hackensack, NJ 07601

UK office: 57 Shelton Street, Covent Garden, London WC2H 9HE

Library of Congress Control Number: 2020941535

British Library Cataloguing-in-Publication Data
A catalogue record for this book is available from the British Library.

LECTURE NOTES ON MECHANICS
Intermediate Level

ISBN 978-981-121-310-6 (hardcover)
ISBN 978-981-121-437-0 (paperback)
ISBN 978-981-121-311-3 (ebook for institutions)
ISBN 978-981-121-312-0 (ebook for individuals)

For any available supplementary material, please visit
https://www.worldscientific.com/worldscibooks/10.1142/11633#t=suppl

Typeset by Stallion Press
Email: enquiries@stallionpress.com

Preface

The primary motivation of this book is to present mechanics at an intermediate level that lies between introductory university text and sophomore classical mechanics. The materials covered are for students having prior experience in mechanics and hence would prefer a treatment that addresses more advanced concepts beyond the scope of a first course in mechanics. As such, the topics encompassed in this book include the co-ordinate free nature of vectors, inertial and non-inertial reference frames, Galilean relativity, Newton's law with variable masses, and rotational dynamics with unconstrained axis-of-rotation in a two-dimensional space.

This book ensued from efforts to meet the intellectual curiosity of students who desire to go beyond the basic concepts of mechanics introduced in standard university textbooks. It is a consequence of the incessant questioning and probing of these earnest students, and through answering their queries that the structure of this book eventually emerged. Although some of the developed materials here are already available in the literature, our book consolidated and placed them within a framework that is fundamental to classical mechanics. In this respect, the contributions of past batches of students who had taken the freshmen course PAP111 Mechanics and Relativity, and its subsequent modified version PH1101/PH1104 Mechanics, are vital and are hereby duly acknowledged. In particular, we would like to express our greatest appreciation to Andri Pradana, who had put in tremendous effort in making the final form of this book possible. Andri had not only created the numerous figures in the book, he had given his time to carefully proof-read and edit the draft manuscript, and by giving many valuable and constructive criticisms. In addition, we are grateful to Suryadi, Lian Jie He, and Matthew Ho for their help in designing the

v

figures of the book. Lastly, we are thankful to our editor Yong Qi Soh from World Scientific, whose active participation and advice are crucial for the final completion of this book.

Lock Yue Chew and Elbert Chia

Contents

Preface   v

1. Vectors as Coordinate-Free System   1

   1.1   Introduction   1
   1.2   Vectors and its Properties   1
       1.2.1   Definition   1
       1.2.2   Basic Algebraic Properties   2
       1.2.3   Vector Multiplication   4
   1.3   Elementary Vector Calculus   6
   1.4   Coordinate Systems   7
       1.4.1   Cartesian Coordinate System   7
       1.4.2   Polar Coordinate System   9

2. Vector Representation of Kinematics   13

   2.1   Introduction   13
   2.2   Spatial Displacement and Time Interval   13
   2.3   Velocity   15
   2.4   Acceleration   17
   2.5   Kinematic Equations   18
   2.6   Applications to Projectile Motion   19
       2.6.1   Projectile Motion over a Horizontal Ground . . .   19
       2.6.2   Projectile Motion over an Inclined Plane   22
   2.7   Kinematics of Circular Motion   26

3. Galilean Theory of Relativity   29

   3.1   Introduction   29
   3.2   Reference Frame   29

3.3   Relative Motion   30
3.4   Examples of Relative Motion   31
    3.4.1   Relative Displacement   31
    3.4.2   Relative Velocity   33
    3.4.3   Relative Acceleration   35
3.5   Galilean Relativity   37

4. Newton's Laws   39

4.1   Introduction   39
4.2   Real Forces   39
4.3   Newton's First Law   40
4.4   Newton's Second Law   41
4.5   Newton's Third Law   42
4.6   Friction and Resistive Forces   44
4.7   Centripetal Force   48
4.8   An Example on the Application of Centripetal Force and
    Static Friction   49
4.9   Non-inertial Reference Frame and the Fictitious Force . .   51
4.10   Examples on the Application of Non-Inertial Reference
    Frame   56
    4.10.1   Object Sliding Up an Accelerating Wedge   56
    4.10.2   Object Sliding Down an Accelerating Wedge . . .   58

5. Energy and Work   61

5.1   Introduction   61
5.2   System and Environment   61
5.3   Work   62
5.4   Hooke's law and the Elastic Spring   65
5.5   Energy Transfer in Spring-Mass System   66
    5.5.1   Work Done by Applied Force on Spring-Mass
    System   66
    5.5.2   Quasi-static Transfer of Energy   67
    5.5.3   Spring-Mass System as an Isolated System   68
    5.5.4   Dissipative Interaction of Spring-Mass System with
    the Environment   70
5.6   Potential Energy and Conservative Force   70
5.7   Collision of Two Identical Mass via a Spring-like Force . .   72
5.8   Pseudowork   76

6. Linear Momentum and Newton's Law with Variable Masses   79

    6.1   Introduction   79
    6.2   Momentum and Newton's Laws of Motion   79
    6.3   Impulsive Force and the Collision Problem   81
    6.4   Elementary Physics of Many-Body Interacting Systems . .   84
          6.4.1   Center of Mass   84
          6.4.2   Mechanics of Many-Body Interacting System and
                  the Conservation of Linear Momentum   85
    6.5   Three Conceptual Examples   87
          6.5.1   Impulsive Projectile Collision and the
                  Conservation of Linear Momentum   87
          6.5.2   Linear Momentum and Kinetic Energy
                  of a System with respect to Different Inertial
                  Reference Frame   89
          6.5.3   Interesting Facts about Elastic Collision   90
    6.6   Newton's Law with Variable Masses   94
          6.6.1   Application to Variable Mass Systems   98

7. The Fundamental Mechanics of Rotational Motion   103

    7.1   Introduction   103
    7.2   Rigid Body and Rotational Kinematics   103
          7.2.1   Rigid Body   103
          7.2.2   Rotational Kinematics   104
    7.3   Rotational Dynamics   108
          7.3.1   Torque and Angular Momentum   109
          7.3.2   Conservation of Angular Momentum   110
          7.3.3   Rotation about a Fixed Axis   111
          7.3.4   Theoretical Framework for Rotational Dynamics
                  with both Translational and Rotational Motion . .   117
          7.3.5   Rolling Motion   126
          7.3.6   Examples on Rolling Motion   129
          7.3.7   Final Examples on Rotational Dynamics   148

Appendix A   Appendix   153

    A.1   Fictitious Forces in a Rotating Reference Frame
          of Constant Angular Velocity   153

Index   159

Chapter 1

Vectors as Coordinate-Free System

1.1 Introduction

The mathematical language of mechanics is that of *vectors*. Thus, a good grasp on the basic concepts of mechanics requires a clear understanding on what vectors are. It is the purpose of this chapter to present the essence of vectors and its diverse mathematical properties.

1.2 Vectors and its Properties

1.2.1 *Definition*

Vectors are quantities defined to have both a magnitude and a direction. Symbolically, it is written as $\vec{v}$ where the overhead arrow indicates direction and its magnitude is represented by $|\vec{v}|$. A graphical representation is shown in Fig. 1.1.

Fig. 1.1

Another mathematical quantity of importance in mechanics is a *scalar*. Unlike vectors, a scalar is just a number which can either be positive or negative — this is what it means when we say that scalar only has magnitude. Later, you will see that vectors can be represented by a group of more than one number.

Overall, scalars and vectors belong to the class of tensors. A scalar is basically the zeroth-order tensor while vectors the first-order tensor.

Higher-order tensors are important in advanced mechanics in the form of the inertial tensor and the stress tensor. Tensor provides a coordinate free representation for quantities defined in physics. As vectors are tensors, they also possess the intrinsic nature of being coordinate free. The meaning of this will become clear in the following development.

1.2.2   Basic Algebraic Properties

First, let us look at the two vectors $\vec{A}$ and $\vec{B}$ displayed in Fig. 1.2. Notice that these two vectors have exactly the same direction and magnitude; however, they act on different points in space. The basic question is, are they equal? Mathematically, $\vec{A} = \vec{B}$. However, in the physical context of mechanics, in which the point of application of a vector does matter[1], $\vec{A}$ and $\vec{B}$ can have different physical effects. In general, every vector has a point of application p and it is from here that the vector begins and extends out in its direction in the form of an arrow (see Fig. 1.2).

Fig. 1.2

Next, let us define the most basic operation on vectors: the *vector addition*. Figure 1.3(a) shows how two vectors $\vec{A}$ and $\vec{B}$ add to give a new vector $\vec{C}$, i.e.

$$\vec{C} = \vec{A} + \vec{B}. \qquad (1.1)$$

Graphically, the addition is performed by putting the tail of $\vec{B}$ on the head of $\vec{A}$, such that $\vec{C}$ extends from the tail of $\vec{A}$ to the head of $\vec{B}$. Such a procedure can be extended to more than three vectors as shown in Fig. 1.3(b):

$$\vec{R} = \vec{A} + \vec{B} + \vec{C} + \vec{D} + \vec{E} + \vec{F} + \vec{G} + \vec{H}. \qquad (1.2)$$

---

[1]For example, the point of application of a force does affect the moment of a force. While moving a vector around in space does not change it mathematically, displacing a vector that represents a physical quantity spatially can have physical consequences. This fact will be illustrated in subsequent chapters.

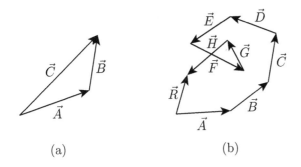

(a)   (b)

Fig. 1.3

An important property of vector addition is the *commutative law of vector addition*:

$$\vec{A} + \vec{B} = \vec{B} + \vec{A},\qquad(1.3)$$

which is easily proven through the construction of a parallelogram as shown in Fig. 1.4. Thus, this law is also known as the parallelogram law.

Fig. 1.4   Fig. 1.5

There is also another law known as the *associative law of vector addition*:

$$\vec{A} + (\vec{B} + \vec{C}) = (\vec{A} + \vec{B}) + \vec{C},\qquad(1.4)$$

whose proof is illustrated in Fig. 1.5.

It is now appropriate for us to introduce the *negative* of a vector. The negative of a vector is a vector of the same magnitude but has a direction directly opposite to the original vector. An illustration of the negative of $\vec{A}$, which is indicated as $-\vec{A}$, is shown in Fig. 1.6. Note that the negative, i.e. $-$, can be viewed as an operator that reverses direction. Thus, $-(-\vec{A}) = \vec{A}$.

The definition of the negative now allows us to define the *zero vector*. It is the vector addition of a vector and its negative, i.e.

$$\vec{A} + (-\vec{A}) = \vec{0}.\qquad(1.5)$$

Fig. 1.6   Fig. 1.7

A zero vector has a magnitude of zero but undefined direction (see Fig. 1.7).

Moreover, the negative allows us to define the operation of *vector subtraction*, as follows:

$$\vec{A} - \vec{B} = \vec{A} + (-\vec{B}).\tag{1.6}$$

The result of vector subtraction is displayed in Fig. 1.8.

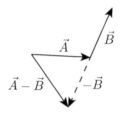

Fig. 1.8

1.2.3   *Vector Multiplication*

Let us next look at *vector multiplication*, which have three varieties. The first is *vector multiplied by a scalar m*, which is expressed as follows:

$$\vec{B} = m\vec{A}.\tag{1.7}$$

Note that if m is positive, $\vec{B}$ has the same direction as $\vec{A}$, but with magnitude $|m\vec{A}|$. Conversely, if m is negative, $\vec{B}$ has a direction that is opposite of $\vec{A}$ and a magnitude of $|m\vec{A}|$.

The second type of vector multiplication is the *dot product*. It is also known as the *scalar product* because its result is a scalar. Its definition is given as follows:

$$\vec{A} \cdot \vec{B} = |\vec{A}||\vec{B}|\cos\theta,\tag{1.8}$$

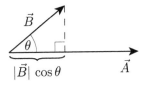

Fig. 1.9

where $|\vec{A}|$ and $|\vec{B}|$ are the magnitude of $\vec{A}$ and $\vec{B}$ respectively, and θ is the smaller angle subtended between $\vec{A}$ and $\vec{B}$ as shown in Fig. 1.9. Geometrically, one can view the dot product as the usual product of the magnitude of $\vec{A}$ and the projection of $\vec{B}$ in the direction of $\vec{A}$, or vice versa. It is obvious from its definition that the dot product is *commutative*:

$$\vec{A} \cdot \vec{B} = \vec{B} \cdot \vec{A}. \tag{1.9}$$

It also obeys the *distributive law of multiplication*:

$$\vec{A} \cdot (\vec{B} + \vec{C}) = \vec{A} \cdot \vec{B} + \vec{A} \cdot \vec{C}, \tag{1.10}$$

which you should prove it for yourself. There are two important facts derived from the dot product. The first is the idea of orthogonal vectors. These are vectors that are perpendicular to each other. For example, if $\vec{A}$ and $\vec{B}$ are orthogonal with each other (see Fig. 1.10), then $\vec{A} \cdot \vec{B} = 0$ (since $\cos 90° = 0$). The second is when the two vectors are parallel or anti-parallel with each other. In the former case, we have $\vec{A} \cdot \vec{B} = |\vec{A}||\vec{B}|$ (since $\cos 0° = 1$), while the latter case $\vec{A} \cdot \vec{B} = -|\vec{A}||\vec{B}|$ (since $\cos 180° = -1$).

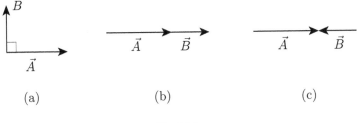

(a)   (b)   (c)

Fig. 1.10

The third type of vector multiplication is known as the *cross product*. It is defined as follows:

$$\vec{A} \times \vec{B} = |\vec{A}||\vec{B}| \sin \theta \, \hat{n}, \tag{1.11}$$

where θ is the smaller subtended angle between $\vec{A}$ and $\vec{B}$, after $\vec{A}$ and $\vec{B}$ have been shifted to the same starting point as shown in Fig. 1.11. $\hat{n}$ is a unit vector which points in a direction determined by the right-hand rule[2], with $\hat{n}$ being normal to the plane spanned by $\vec{A}$ and $\vec{B}$. Thus, the result of a cross product is a vector whose magnitude is the area of the parallelogram formed by $\vec{A}$ and $\vec{B}$ (see Fig. 1.11). Finally, note that

$$\vec{A} \times \vec{B} = -\vec{B} \times \vec{A}. \tag{1.12}$$

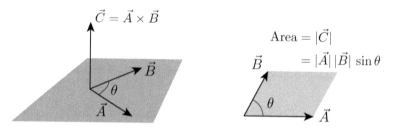

Fig. 1.11

1.3 Elementary Vector Calculus

The following is a list of vector differentiation formulas that will be useful for the development of our formalism of mechanics in subsequent chapters.

(1) $\dfrac{d}{dt}(\vec{A} + \vec{B}) = \dfrac{d\vec{A}}{dt} + \dfrac{d\vec{B}}{dt}$

(2) $\dfrac{d}{dt}(m\vec{A}) = \dfrac{dm}{dt}\vec{A} + m\dfrac{d\vec{A}}{dt}$

(3) $\dfrac{d}{dt}(\vec{A} \cdot \vec{B}) = \dfrac{d\vec{A}}{dt} \cdot \vec{B} + \vec{A} \cdot \dfrac{d\vec{B}}{dt}$

(4) $\dfrac{d}{dt}(\vec{A} \times \vec{B}) = \dfrac{d\vec{A}}{dt} \times \vec{B} + \vec{A} \times \dfrac{d\vec{B}}{dt}$

(5) $\vec{B} = \dfrac{d\vec{A}}{dt} \implies d\vec{A} = \vec{B}dt \implies \displaystyle\int_{\vec{A}_i}^{\vec{A}_f} d\vec{A} = \int_{t_i}^{t_f} \vec{B}dt$

   $\implies \vec{A}_f - \vec{A}_i = \displaystyle\int_{t_i}^{t_f} \vec{B}dt$

---

[2]The rule states that the fingers of the right hand sweep from $\vec{A}$ to $\vec{B}$ and the thumb then gives the direction of $\hat{n}$

1.4   Coordinate Systems

So far, vectors have been presented in a coordinate-free manner, i.e. no coordinate systems have been introduced for its description. Vectors in this form is succinct and has the power of highlighting the main physical quantities involved, such as velocity, acceleration, force, momentum etc. that will be introduced later. However, when solving a mechanical problem, it becomes necessary to define a coordinate system. Nonetheless, the fact that vectors are coordinate-free allows one the freedom to choose the appropriate coordinate system that simplifies the solution of the problem.

1.4.1   *Cartesian Coordinate System*

The most versatile and popular coordinate system is the Cartesian coordinate system. It is a system where a point in space is located by its perpendicular distance to three-axes (the x-y-z axes for 3-dimensional space) which are orthogonal to each other (see Fig. 1.12 for a 2-dimensional version). For the 2D case, the axes are x and y. Associated with each axis is their respective basis vector: $\hat{\imath}$ for the x axis and $\hat{\jmath}$ for the y axis. Note that a basis vector points in the direction of increasing coordinate. Moreover, since there are two coordinates in the 2D plane, we expect two basis vectors at each point in the plane. Thus, one can view the set of basis vectors to define a vector field in the plane (see Fig. 1.12). In the case of Cartesian coordinates, they are unit vectors at each point such that

$$\hat{\imath} \cdot \hat{\imath} = \hat{\jmath} \cdot \hat{\jmath} = |\hat{\imath}| = |\hat{\jmath}| = 1 \,. \tag{1.13}$$

In addition, these vectors are orthogonal, i.e.,

$$\hat{\imath} \cdot \hat{\jmath} = 0 \,. \tag{1.14}$$

It is important to note that the $\hat{\imath}$ and $\hat{\jmath}$ vector has the same magnitude and direction at all x and y coordinates, i.e. they are independent of (x, y) (see Fig. 1.12).

Figure 1.13 shows an attempt to represent the vector $\vec{A}$ with respect to a given Cartesian coordinate system. In this system, $\vec{A}$ makes an angle θ_y and θ_x with the y-axis and x-axis respectively. Then, the projection of $\vec{A}$ to the y-axis is $A_y = \vec{A} \cdot \hat{\jmath} = |\vec{A}| \cos \theta_y$, while that to the x-axis is $A_x = \vec{A} \cdot \hat{\imath} = |\vec{A}| \cos \theta_x$. With respect to this Cartesian coordinate system,

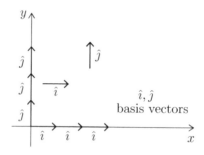

Fig. 1.12

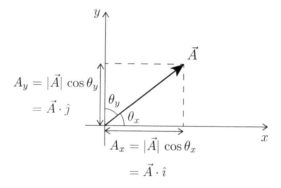

Fig. 1.13

$\vec{A}$ can be written in terms of the basis vectors as follows:

$$\vec{A} = A_x \hat{\imath} + A_y \hat{\jmath}$$
$$= \left(\vec{A} \cdot \hat{\imath}\right) \hat{\imath} + \left(\vec{A} \cdot \hat{\jmath}\right) \hat{\jmath}$$
$$= A_x \begin{pmatrix} 1 \\ 0 \end{pmatrix} + A_y \begin{pmatrix} 0 \\ 1 \end{pmatrix}$$
$$= \begin{pmatrix} A_x \\ A_y \end{pmatrix}_{x-y}, \qquad (1.15)$$

where in the last line of the equation, we have expressed $\vec{A}$ in the form of $\hat{\imath} = (1\ 0)^T$ and $\hat{\jmath} = (0\ 1)^T$. Note that the superscript T represents the transpose operation.

Since $\vec{A}$ is coordinate-free, we can express this same vector in another coordinate system as shown in Fig. 1.14. In this system, $\vec{A}$ lies entirely

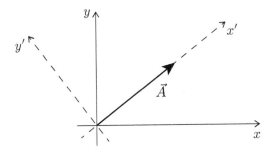

Fig. 1.14

in the x' axis, and its decomposition in terms of the corresponding basis vector $\hat{\imath}'$ and $\hat{\jmath}'$ is

$$\vec{A} = \sqrt{A_x^2 + A_y^2}\ \hat{\imath}'$$

$$= \sqrt{A_x^2 + A_y^2} \begin{pmatrix} 1 \\ 0 \end{pmatrix}$$

$$= \begin{pmatrix} \sqrt{A_x^2 + A_y^2} \\ 0 \end{pmatrix}_{x'-y'} . \tag{1.16}$$

The choice of the coordinate system depends on the one that is best suited to solve the problem.

1.4.2 *Polar Coordinate System*

For problems involving circular symmetry, for example circular motion, the polar coordinate system is the more suitable system to represent the vectors in the problem.

The polar coordinate system is defined as shown in Fig. 1.15, where the location of the point is determined by its radial distance r to the origin, and θ is the angle subtended by the radial line from the point to the origin with the positive x-axis. The mathematical transformation between the polar coordinates and the Cartesian coordinates is given by the following set of expressions:

$$x = r\cos\theta, \tag{1.17}$$

$$y = r\sin\theta \tag{1.18}$$

and

$$r = \sqrt{x^2 + y^2}, \tag{1.19}$$

$$\theta = \arctan\left(\frac{y}{x}\right). \tag{1.20}$$

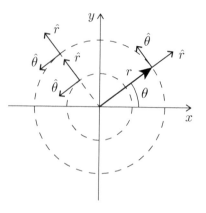

Fig. 1.15

As in the Cartesian coordinate system, basis vectors can be defined to be pointing in the direction of increasing coordinate, and we expect two basis vectors per point in the plane. From this perspective, we can deduce that the unit vectors that correspond to the polar coordinate system are (see Fig. 1.15): $\hat{r}$ and $\hat{\theta}$, i.e. $\hat{r} \cdot \hat{r} = \hat{\theta} \cdot \hat{\theta} = |\hat{r}| = |\hat{\theta}| = 1$. Note that $\hat{r}$ and $\hat{\theta}$ are orthogonal, or $\hat{r} \cdot \hat{\theta} = 0$. Like the basis vectors in the Cartesian coordinate system, the collection of $\hat{r}$ and $\hat{\theta}$ define a field of basis vectors. But unlike the Cartesian coordinate system, each $\hat{r}$ and $\hat{\theta}$ depends on its spatial position in the polar coordinate system. Specifically, $\hat{r}$ (and $\hat{\theta}$) has different direction at different spatial position (see Fig. 1.15). This fact is made obvious by relating $\hat{r}$ and $\hat{\theta}$ to the Cartesian coordinate basis $\hat{\imath}$ and $\hat{\jmath}$:

$$\hat{r} = \cos\theta\hat{\imath} + \sin\theta\hat{\jmath}, \tag{1.21}$$
$$\hat{\theta} = -\sin\theta\hat{\imath} + \cos\theta\hat{\jmath}. \tag{1.22}$$

Figure 1.16 illustrates how Eqs. (1.21) and (1.22) are derived. Thus, the specific direction of the basis vectors of the polar coordinate system do depend on space, i.e., they are implicitly dependent on the angular displacement from the positive x-axis. While $\hat{r}$ and $\hat{\theta}$ can be defined on the spatial plane based on Eqs. (1.21) and (1.22), they are however undefined at one point in this plane, that is the origin. At this point, the value of θ is undefined, and hence $\hat{r}$ and $\hat{\theta}$ are undefined as a result of Eqs. (1.21) and (1.22).

Finally, the usefulness of such a representation is exemplified by two examples in mechanics where circular symmetry (or more specifically, spherical symmetry) is observed. The first is the vectorial representation of the

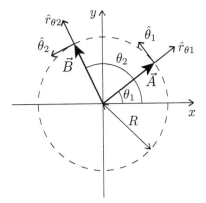

Fig. 1.16

gravitational force (see Fig. 1.17):

$$\vec{F}_G = -\frac{GMm}{R^2}\hat{r},$$ (1.23)

and the other is that of centripetal acceleration (see Fig. 1.18):

$$\vec{a}_c = -\frac{v^2}{R}\hat{r}.$$ (1.24)

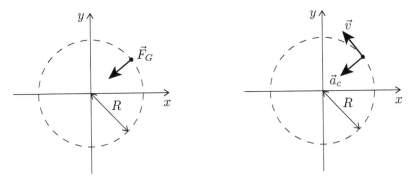

Fig. 1.17   Fig. 1.18

In summary, a set of basis vectors is defined with respect to its coordinate system. The above discussion shows that one must always be careful to interpret the meaning of the basis vectors in the context of its coordinate system.

Chapter 2

Vector Representation of Kinematics

2.1 Introduction

The study of motion begins with kinematics. Here, we start to define physical quantities that are measurable. From these quantities, more complex and even abstract quantities can be built. In this chapter, we associate these physical quantities to the vectors and the vector algebra covered in the last chapter. Moreover, it is important to note that these basic quantities are intimately related to the fundamental notion of space and time.

2.2 Spatial Displacement and Time Interval

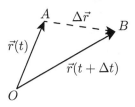

Fig. 2.1

The first two quantities of interest are spatial displacement (or *displacement* in short) and *time interval*. The best way to appreciate these quantities is to consider the motion of a particle [1] from a point A in space, to another point B in space. Consider an observer O as shown in Fig. 2.1. With respect to O, the *position* of the particle at A at *time instant* t can

---

[1]Here, particle is a way for us to represent an arbitrary object. Since we are concerned with linear translation in the earlier chapters, the particle can also be viewed as a representation of the center of mass of the object.

be described by the *position vector* $\vec{r}(t)$. After a time interval Δt, that is
at a later time instant $t + \Delta t$, O observes that the particle has moved to B
whose location is given by $\vec{r}(t + \Delta t)$. The displacement of the particle $\Delta \vec{r}$
is given by the difference between final position vector $\vec{r}(t + \Delta t)$ and initial
position vector $\vec{r}(t)$, that is

$$\Delta \vec{r} = \vec{r}(t + \Delta t) - \vec{r}(t). \tag{2.1}$$

Thus, the motion has caused the particle to *displace* spatially from point
A to point B. Such a spatial displacement can be represented by a vector
$\Delta \vec{r}$, with the tail of the arrow at A and its head at B (see also Fig. 2.2
for another example). This vector gives both the direction and magnitude
of the displacement. It is important to note that displacement is not an
abstract mathematical quantity. It is measurable, and hence physical. In
terms of SI units, the unit of measurement is the *meter*.

Fig. 2.2

Displacement is a vector quantity and follows the algebra of vectors.
For example, if a particle first moves from A to B by a displacement $\Delta \vec{r}_1$,
and then from B to C by a displacement $\Delta \vec{r}_2$, the resultant displacement
of the overall motion is given by $\Delta \vec{R}$ as shown in Fig. 2.3, where

$$\Delta \vec{R} = \Delta \vec{r}_1 + \Delta \vec{r}_2, \tag{2.2}$$

which is a vector addition.

Fig. 2.3   Fig. 2.4

Now, if the particle continues to move from C back to A, such as to
complete a round trip as shown in Fig. 2.4, we observe that the total
displacement

$$\Delta \vec{r}_1 + \Delta \vec{r}_2 + \Delta \vec{r}_3 = 0. \tag{2.3}$$

The resultant displacement of the particle is thus zero. At this juncture, let us introduce another measurable quantity known as *distance*. The distance is defined as the magnitude of each displacement. In the example of Fig. 2.4, we have the distance

$$d_i = |\Delta \vec{r}_i| \quad i = 1, 2, 3 \,. \tag{2.4}$$

Unlike displacement, distance is a scalar physical quantity. Hence, it obeys the algebra of scalars. Specifically, we have that total distance, $d_{\text{total}} = d_1 + d_2 + d_3$.

Similarly, time is a scalar and obeys scalar algebra. For example, let the time taken for the particle to move from A to B in Fig. 2.3 be Δt_1, which represents the time interval between the events. Similar descriptions are to be taken for the time interval Δt_2 from B to C, and Δt_3 from C to A. By applying the notion that time is a scalar, we deduce the time interval between event A and event C to be $\Delta t_1 + \Delta t_2$. In addition, the time taken for the particle to move from event A to B to C and then back to A is $\Delta t_1 + \Delta t_2 + \Delta t_3$, which is analogous to the total distance discussed earlier.

It is important to note that there are certain notions that cannot be made definite as displacement and time. Let us take the example of the color "green". If we were to try to define the color "green" as a physical quantity, we fail because

$$\text{green} + \text{green} = \text{green} \neq 2 \, \text{green} \,.$$

The notion "green" defined in this sense is not a measurable quantity.

2.3   Velocity

With the definition of displacement Δr and time interval Δt in the last two sections, it is appropriate now for us to use these two concepts to construct a more complex physical quantity: *velocity*.

For this, let us suppose that two particles displace from A to B as in Fig. 2.1 but they take two different time intervals. Thus, one particle is moving faster than the other. This leads to the definition of average velocity as follows:

$$\vec{v}_{av} = \frac{\Delta \vec{r}}{\Delta t} \,, \tag{2.5}$$

which shows that velocity is a vector with a unit of meter per second. It is instructive to take another look at Eq. (2.5) with respect to Fig. 2.1 and Eq. (2.1). In terms of the position vectors there, Eq. (2.5) becomes

$$\vec{v}_{av} = \frac{\vec{r}(t + \Delta t) - \vec{r}(t)}{\Delta t} \,. \tag{2.6}$$

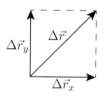

Fig. 2.5

Now, if a person were to move due north on the deck of a boat by $\Delta\vec{r}_y$ while the boat moves due east by $\Delta\vec{r}_x$ simultaneously within a time interval of Δt, the resultant displacement of the person is

$$\Delta\vec{r} = \Delta\vec{r}_x + \Delta\vec{r}_y \tag{2.7}$$

within Δt (see Fig. 2.5) relative to the ground. By considering the rate of displacement of the person, the boat, and the resultant displacement, through scaling Eq. (2.7) by $1/\Delta t$, we obtain

$$\frac{\Delta\vec{r}}{\Delta t} = \frac{\Delta\vec{r}_x}{\Delta t} + \frac{\Delta\vec{r}_y}{\Delta t}, \tag{2.8}$$

which shows that the resultant velocity obeys the law of vector addition:

$$\vec{v}_{av} = \vec{v}_{av_x} + \vec{v}_{av_y}. \tag{2.9}$$

Note that this result is true in general, which can be observed from the fact that

$$\vec{v}_{av} = \vec{v}_{av_1} + \vec{v}_{av_2} \tag{2.10}$$

as shown in Fig. 2.6(a), is a scaled version (scaled by $1/\Delta t$) of the corresponding displacement vector addition diagram shown in Fig. 2.6(b).

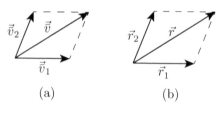

(a)   (b)

Fig. 2.6

In the limit that the time interval at which the displacement occurs is infinitesimally small, i.e. $\Delta t \to 0$, we have the *instantaneous* velocity, defined as:

$$\vec{v} = \lim_{\Delta t \to 0} \frac{\Delta \vec{r}}{\Delta t} = \frac{d\vec{r}}{dt}. \tag{2.11}$$

In this sense, velocity is defined as the instantaneous rate of displacement. Based on the above argument, it is easy to see that velocity obeys:

$$\vec{v} = \vec{v}_1 + \vec{v}_2. \tag{2.12}$$

2.4   Acceleration

The last kinematic quantity that we are going to construct is acceleration. The average acceleration is defined as

$$\vec{a}_{av} = \frac{\Delta \vec{v}}{\Delta t}, \tag{2.13}$$

where

$$\Delta \vec{v} = \vec{v}(t + \Delta t) - \vec{v}(t) \tag{2.14}$$

is the change in velocity within a time interval Δt. Acceleration has a unit of meter per second squared and it is also a vector. Therefore, it obeys the law of vector addition:

$$\vec{a}_{av} = \vec{a}_{av_1} + \vec{a}_{av_2}. \tag{2.15}$$

Let us give a physical illustration of Eq. (2.15). Consider the example of the person walking on the moving boat again. If the velocity of the person changes by $\Delta \vec{v}_y$, and the velocity of the boat changes by $\Delta \vec{v}_x$ within the same time interval Δt, the resultant change in velocity of the person is

$$\Delta \vec{v} = \Delta \vec{v}_x + \Delta \vec{v}_y. \tag{2.16}$$

Since these changes are simultaneous, we can scale Eq. (2.16) by $1/\Delta t$ just like Eq. (2.8):

$$\frac{\Delta \vec{v}}{\Delta t} = \frac{\Delta \vec{v}_x}{\Delta t} + \frac{\Delta \vec{v}_y}{\Delta t}, \tag{2.17}$$

giving

$$\vec{a}_{av} = \vec{a}_{av_x} + \vec{a}_{av_y}. \tag{2.18}$$

Similarly, in the infinitesimal time interval $\Delta t \to 0$, we get the instantaneous acceleration

$$\vec{a} = \lim_{\Delta t \to 0} \frac{\Delta \vec{v}}{\Delta t} = \frac{d\vec{v}}{dt} = \frac{d^2\vec{r}}{dt^2}, \tag{2.19}$$

which relates a change in velocity with infinitesimal time interval. Again, the following holds:

$$\vec{a} = \vec{a}_1 + \vec{a}_2. \tag{2.20}$$

2.5   Kinematic Equations

Let us consolidate the kinematic equations obtained in the last two sections:

$$\vec{a} = \frac{d\vec{v}}{dt} \, , \tag{2.21}$$

$$\vec{v} = \frac{d\vec{r}}{dt} \, . \tag{2.22}$$

These equations are the definition of acceleration and velocity in the differential form respectively. They can be transformed into equations in the integral form using property (5) of the elementary vector calculus discussed in the last chapter:

$$\vec{v}_f = \vec{v}_i + \int_{t_i}^{t_f} \vec{a} \, dt \, , \tag{2.23}$$

$$\vec{r}_f = \vec{r}_i + \int_{t_i}^{t_f} \vec{v} \, dt \, . \tag{2.24}$$

In general, the evaluation of Eqs. (2.23) and (2.24) requires a precise specification of the functional dependence of $\vec{a}$ and $\vec{v}$ against time t. Thus, the possibility of solving each equation analytically depends on the corresponding functional form. In particular, when the acceleration $\vec{a}$ is a constant in time, such analytical solutions can be obtained.

Next we work with the special case where $\vec{a}$ is a constant vector. Applying this condition to Eq. (2.23), we have

$$\vec{v}(t) = \vec{v}_i + \int_{t_i}^{t} \vec{a} \, dt' \, ,$$

$$= \vec{v}_i + \vec{a} \int_{t_i}^{t} dt' \, ,$$

$$= \vec{v}_i + \vec{a} \, (t - t_i) \, ,$$

where we have assumed $\vec{v}_i$ to be the velocity of the particle at time $t = t_i$ and $\vec{v}$ is the velocity at time t. Reiterating, we have

$$\vec{v}(t) = \vec{v}_i + \vec{a} \, (t - t_i) \, , \tag{2.25}$$

which is an expression of the velocity of the particle at time instant t of its *constant* accelerating motion. Substituting this expression into Eq. (2.24), we have

$$\vec{r}(t) = \vec{r}_i + \int_{t_i}^{t} \vec{v} \, dt' \, ,$$

$$= \vec{r}_i + \int_{t_i}^{t} \left[\vec{v}_i + \vec{a} \left(t' - t_i \right) \right] dt' \, ,$$

$$= \vec{r}_i + \int_{t_i}^{t} \vec{v}_i \, dt' + \int_{t_i}^{t} \vec{a} \left(t' - t_i \right) dt' \, ,$$

$$= \vec{r}_i + \vec{v}_i \int_{t_i}^{t} dt' + \vec{a} \int_{t_i}^{t} \left(t' - t_i \right) dt' \, ,$$

$$= \vec{r}_i + \vec{v}_i \left(t - t_i \right) + \vec{a} \frac{\left(t - t_i \right)^2}{2} \, ,$$

Thus, we have

$$\vec{r}(t) = \vec{r}_i + \vec{v}_i \left(t - t_i \right) + \frac{1}{2} \vec{a} \left(t - t_i \right)^2 \, , \tag{2.26}$$

where $\vec{r}_i$ is the position of the particle at time $t = t_i$ and $\vec{r}$ its position at time t.

2.6 Applications to Projectile Motion

In this section, we will use a geometric approach based on the vector kinematic equations (2.25) and (2.26) to solve two problems relating to projectile motions, different from the usual algebraic approach in typical freshman textbooks. In both cases, we shall assume that $t_i = 0$ and $r_i = 0$. Also, $\vec{a} = \vec{g}$ is the gravitational acceleration which is directing vertically downward.

2.6.1 *Projectile Motion over a Horizontal Ground*

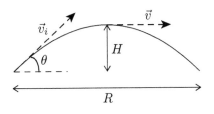

Fig. 2.7

The path of the projectile is illustrated in Fig. 2.7 for the first problem. Let us first determine the time t_H at which the projectile is at its maximum height. At its maximum height, the velocity $\vec{v}(t_H)$ of the projectile is horizontal. Since the initial velocity of the projectile $\vec{v}_i$ is inclined at an angle of θ to the horizontal, we can draw a vector triangle based on Eq. (2.25) as shown in Fig. 2.8 with $\vec{v}$ being the velocity at maximum height. This vector triangle allows us to obtain t_H immediately as follows:

$$\sin\theta = \frac{|\vec{g}\, t_H|}{|\vec{v}_i|} = \frac{g\, t_H}{v_i}\,,$$

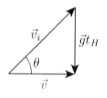

Fig. 2.8

Therefore,

$$t_H = \frac{v_i \sin\theta}{g}\,. \tag{2.27}$$

Next, we shall obtain the maximum height H of the projectile. For this, we shall employ the vector triangle OAB illustrated in Fig. 2.9 which is based on Eq. (2.26) with $t = t_H$. The geometry of the triangle OAC then enables us to perform the following evaluation:

$$\sin\theta = \frac{H_T}{|\vec{v}_i\, t_H|} = \frac{H_T}{v_i\, t_H}\,,$$

where H_T is the height of point A from the ground at t_H if there is no gravity. Using Eq. (2.27), we obtain

$$H_T = v_i \sin\theta \left(\frac{v_i \sin\theta}{g} \right) = \frac{v_i^2 \sin^2\theta}{g}\,.$$

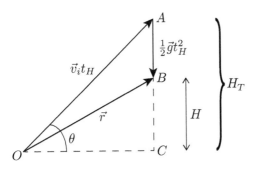

Fig. 2.9

Then, the maximum height H is given by

$$H = H_T - \frac{1}{2}|\vec{g}t_H^2|,$$

$$= H_T - \frac{1}{2}g\,t_H^2,$$

$$= H_T - \frac{1}{2}g\,\frac{v_i^2\sin^2\theta}{g^2},$$

$$= \frac{v_i^2\sin^2\theta}{g} - \frac{v_i^2\sin^2\theta}{2g},$$

$$= \frac{v_i^2\sin^2\theta}{2g}. \tag{2.28}$$

The final quantity to determine is the range R of the projectile. Let t_R be the time instant at which the projectile traverses R. Since the position vector $\vec{R}$ of the point at which the projectile reaches the ground is horizontal, using Eq. (2.26), we can construct a vector triangle as shown in Fig. 2.10. This triangle enables us to perform the following calculations:

$$\sin\theta = \frac{|\frac{1}{2}\vec{g}t_R^2|}{|\vec{v}_i\,t_R|} = \frac{\frac{1}{2}g\,t_R^2}{v_i\,t_R}.$$

Therefore,

$$t_R = \frac{2v_i\sin\theta}{g}. \tag{2.29}$$

Also,

$$\cos\theta = \frac{|\vec{R}|}{|\vec{v}_i\,t_R|} = \frac{R}{v_i\,t_R}.$$

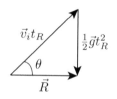

Fig. 2.10

Then, using Eq. (2.29),

$$R = v_i t_R \cos\theta \,,$$

$$= v_i \cos\theta \left(\frac{2v_i \sin\theta}{g} \right) \,,$$

$$= \frac{2v_i^2 \sin\theta \cos\theta}{g} \,.$$

Thus,

$$R = \frac{v_i^2 \sin 2\theta}{g} \,. \tag{2.30}$$

2.6.2 *Projectile Motion over an Inclined Plane*

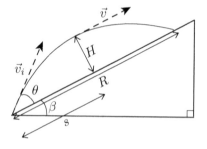

Fig. 2.11

The path of the projectile over an inclined plane is shown in Fig. 2.11. Here, we are to determine (i) the range R, which is the distance from the origin of the projectile to its destination along the inclined plane, (ii) the maximum height H of the projectile over the inclined plane, and (iii) the distance s along the inclined plane from the origin to the point where the maximum height happens. Let us first determine R. Assuming that the projectile reaches its destination at time t_R, we can draw a vector triangle for this case based on Eq. (2.26) as shown in Fig. 2.12. Taking note of the geometry of this problem and applying the sine rule to this triangle, we obtain t_R as follows:

$$\frac{\left|\frac{1}{2}\vec{g}\,t_R^2\right|}{\sin\theta} = \frac{\left|\vec{v}_i\,t_R\right|}{\sin\left(\frac{\pi}{2}+\beta\right)},$$

$$\frac{1}{2}g\,t_R^2\cos\beta = v_i\,t_R\sin\theta,$$

which leads to

$$t_R = \frac{2v_i\sin\theta}{g\cos\beta}. \tag{2.31}$$

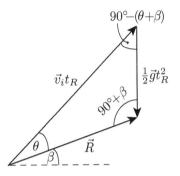

Fig. 2.12

Applying the sine rule one more time to this vector triangle and using Eq. (2.31), we have

$$\frac{\left|\vec{R}\right|}{\sin\left(\frac{\pi}{2}-(\theta+\beta)\right)} = \frac{\left|\vec{v}_i\,t_R\right|}{\sin\left(\frac{\pi}{2}+\beta\right)},$$

$$= \frac{v_i\,t_R}{\cos\beta},$$

$$= \left(\frac{v_i}{\cos\beta}\right)\left(\frac{2v_i\sin\theta}{g\cos\beta}\right),$$

$$= \frac{2v_i^2\sin\theta}{g\cos^2\beta}.$$

Therefore,

$$R = \frac{2v_i^2\sin\theta\cos\left(\theta+\beta\right)}{g\cos^2\beta}. \tag{2.32}$$

Next, we determine the maximum height H of the projectile over the inclined plane. Let t_H be the time at which the projectile reaches this

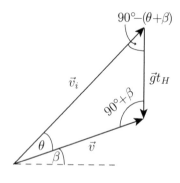

Fig. 2.13

maximum height. Applying the sine rule to the vector triangle which is based on Eq. (2.25) as shown in Fig. 2.13:

$$\frac{|\vec{g}\,t_H|}{\sin\theta} = \frac{|\vec{v}_i|}{\sin\left(\frac{\pi}{2} + \beta\right)},$$

$$\frac{g\,t_H}{\sin\theta} = \frac{v_i}{\cos\beta},$$

we obtain

$$t_H = \frac{v_i \sin\theta}{g \cos\beta}. \tag{2.33}$$

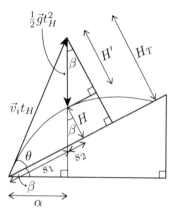

Fig. 2.14

According to Eq. (2.26), we plot the geometry of the situation as illustrated in Fig. 2.14. This geometry will be useful for the evaluation of H

(and also s). From Fig. 2.14, we observe that

$$H_T = v_i t_H \sin\theta,$$ (2.34)

and

$$H' = \frac{1}{2}g t_H^2 \cos\beta.$$ (2.35)

To obtain H, we subtract Eq. (2.34) from Eq. (2.35):

$$H = H_T - H',$$
$$= v_i t_H \sin\theta - \frac{1}{2}g t_H^2 \cos\beta,$$
$$= v_i \sin\theta \left(\frac{v_i \sin\theta}{g\cos\beta}\right) - \frac{1}{2}g\cos\beta \left(\frac{v_i^2 \sin^2\theta}{g^2\cos^2\beta}\right),$$
$$= \frac{v_i^2 \sin^2\theta}{g\cos\beta} - \frac{v_i^2 \sin^2\theta}{2g\cos\beta},$$

which leads to

$$H = \frac{v_i^2 \sin^2\theta}{2g\cos\beta}.$$ (2.36)

Finally, we proceed to evaluate the last quantity of interest, which is the distance s to the maximum point along the inclined plane. As an intermediate step, we yield the distance d as follows:

$$d = v_i t_H \cos(\theta + \beta),$$
$$= v_i \cos(\theta + \beta)\left(\frac{v_i \sin\theta}{g\cos\beta}\right),$$
$$= \frac{v_i^2 \sin\theta \cos(\theta + \beta)}{g\cos\beta},$$

which allows us to obtain the distance s_1 (see Fig. 2.14):

$$s_1 = \frac{d}{\cos\beta} = \frac{v_i^2 \sin\theta \cos(\theta + \beta)}{g\cos^2\beta}.$$ (2.37)

Through the geometry of Fig. 2.14, we then determine the distance s_2:

$$s_2 = H\tan\beta = \left(\frac{v_i^2 \sin^2\theta}{2g\cos\beta}\right)\left(\frac{\sin\beta}{\cos\beta}\right) = \frac{v_i^2 \sin^2\theta \sin\beta}{2g\cos^2\beta}.$$ (2.38)

The quantity-of-interest s is the sum of s_1 and s_2, whose evaluation

$$
\begin{aligned}
s &= s_1 + s_2 \,, \\
&= \frac{v_i^2 \sin \theta \cos (\theta + \beta)}{g \cos^2 \beta} + \frac{v_i^2 \sin^2 \theta \sin \beta}{2g \cos^2 \beta}
\end{aligned}
\tag{2.39}
$$

eventually leads to

$$
s = \frac{R}{2} + \frac{v_i^2 \sin^2 \theta \sin \beta}{2g \cos^2 \beta} \,.
\tag{2.40}
$$

2.7 Kinematics of Circular Motion

In this section, we apply the definition of velocity and acceleration to develop the kinematics of a particle executing circular motion using polar coordinates. The motion of the particle is about a circle of radius R and its position vector is given by

$$
\vec{r} = R\hat{r} \,.
\tag{2.41}
$$

We then obtain its velocity $\vec{v}$ as follows:

$$
\begin{aligned}
\vec{v} &= \frac{d\vec{r}}{dt} \,, \\
&= \frac{d\,(R\hat{r})}{dt} \,, \\
&= R\frac{d\hat{r}}{dt} \,.
\end{aligned}
\tag{2.42}
$$

Note that, since the velocity is always tangential to the path, v is purely tangential and does not have a radial component. Because $\hat{r} = \cos \theta \hat{i} + \sin \theta \hat{j}$ according to Eq. (1.21), we deduce that

$$
\begin{aligned}
\frac{d\hat{r}}{dt} &= \frac{d\,(\cos \theta)}{dt}\hat{i} + \cos \theta \frac{d\hat{i}}{dt} + \frac{d\,(\sin \theta)}{dt}\hat{j} + \sin \theta \frac{d\hat{j}}{dt} \\
&= \frac{d\,(\cos \theta)}{dt}\hat{i} + \frac{d\,(\sin \theta)}{dt}\hat{j} \\
&= -\sin \theta \frac{d\theta}{dt}\hat{i} + \cos \theta \frac{d\theta}{dt}\hat{j} \\
&= \omega\,(-\sin \theta \hat{i} + \cos \theta \hat{j}) \\
&= \omega \hat{\theta} \,,
\end{aligned}
\tag{2.43}
$$

where the last line of the above equation is obtained from Eq. (1.22). Furthermore, we have defined the angular velocity $\omega = d\theta/dt$ and made use of the fact that $d\hat{\imath}/dt = d\hat{\jmath}/dt = 0$. Combining Eqs. (2.42) and (2.43), the velocity of the particle is

$$\vec{v} = R\omega\hat{\theta}\,, \tag{2.44}$$

$$= v\hat{\theta}\,, \tag{2.45}$$

where we define the speed v of the particle as

$$v = R\omega\,. \tag{2.46}$$

Equation (2.45) gives the tangential velocity of the particle as it executes circular motion. Notice that for circular motion, the component of the velocity of the particle in the radial direction $\hat{r}$ (i.e., the particle's radial velocity) is zero.

The acceleration $\vec{a}$ of the particle is then given by

$$\vec{a} = \frac{d\vec{v}}{dt}\,,$$

$$= \frac{d\left(v\hat{\theta}\right)}{dt}\,,$$

$$= \frac{dv}{dt}\hat{\theta} + v\frac{d\hat{\theta}}{dt}\,. \tag{2.47}$$

Since $\hat{\theta} = -\sin\theta\hat{\imath} + \cos\theta\hat{\jmath}$ based on Eq. (1.22), we have

$$\frac{d\hat{\theta}}{dt} = \frac{d\left(-\sin\theta\right)}{dt}\hat{\imath} - \sin\theta\frac{d\hat{\imath}}{dt} + \frac{d\left(\cos\theta\right)}{dt}\hat{\jmath} + \cos\theta\frac{d\hat{\jmath}}{dt}$$

$$= \frac{d\left(-\sin\theta\right)}{dt}\hat{\imath} + \frac{d\left(\cos\theta\right)}{dt}\hat{\jmath}$$

$$= -\cos\theta\frac{d\theta}{dt}\hat{\imath} - \sin\theta\frac{d\theta}{dt}\hat{\jmath}$$

$$= -\omega\left(\cos\theta\hat{\imath} + \sin\theta\hat{\jmath}\right)$$

$$= -\omega\hat{r}\,, \tag{2.48}$$

where we have employed Eq. (1.21) to get the last line of the above equation. Noting that $\omega = v/R$, we have

$$\frac{d\hat{\theta}}{dt} = -\frac{v}{R}\hat{r}\,. \tag{2.49}$$

Substituting this equation into Eq. (2.47), we obtain

$$\vec{a} = \frac{dv}{dt}\hat{\theta} + v\left(-\frac{v}{R}\right)\hat{r}$$
$$= \frac{dv}{dt}\hat{\theta} - \left(\frac{v^2}{R}\right)\hat{r}$$
$$= a_t\hat{\theta} + a_r\hat{r}\,. \tag{2.50}$$

This result implies that in circular motion, there are two parts to the acceleration. The first part is *tangential acceleration*:

$$a_t\hat{\theta} = \frac{dv}{dt}\hat{\theta} \tag{2.51}$$

and the second part is *centripetal* (or *radial*) *acceleration*:

$$a_r\hat{r} = -\left(\frac{v^2}{R}\right)\hat{r}\,. \tag{2.52}$$

The decomposition of the acceleration of a particle in circular motion to its tangential and centripetal component can also be seen in Fig. 2.15.

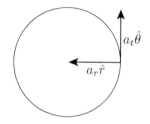

Fig. 2.15

Chapter 3

Galilean Theory of Relativity

3.1 Introduction

In the last chapter, we have defined physical quantities in terms of vectors and have begun the study of motion through the definition of kinematic quantities such as the displacement, velocity, and acceleration. It is important to note that the measurement of these quantities in the last chapter is performed by an observer fixed to a particular reference frame. Since there can be many observers in a physical system and the observers can also be in motion relative to each other, we need to consider the relative motion of the reference frame that is attached to each of these observers.

3.2 Reference Frame

We consider a reference frame to be a point of reference for an observer. It is a frame where physical events occur with respect to this observer. The physical events could be the motion of a ball, collision of two particles, or the explosion of a supernova. An observer O can always describe the events in terms of its space-time coordinates (x, y, z, t). As such, O can describe both the evolution of the events in space and its history in time. Analogously, another observer O' who is in motion relative to O can also observe the same phenomena observed by O. However, O' will describe these same events based on his reference frame with space-time coordinates (x', y', z', t'). It is interesting that O and O' give different accounts of their observation of the same events, which are nonetheless intimately connected through transformation rules of relative motion that will be discussed in the next section.

3.3   Relative Motion

Let us consider two observers O and O' moving relative to each other. In other words, the reference frames S and S' which are attached to O and O' respectively are in relative motion. Pictorially, this is shown in Fig. 3.1. Assume the occurrence of an event A. According to observer O, event A occurs at the spatial location indicated by the position vector $\vec{r}_{A/O}$. Note that the notation "A/O" means A relative to (or with respect to) O. On the other hand, the same event A occurs at the spatial position given by the vector $\vec{r}_{A/O'}$ for observer O'. If the position of O' is at $\vec{r}_{O'/O}$ relative to O, then the following vector addition rule applies (see Fig. 3.1):

$$\vec{r}_{A/O} = \vec{r}_{A/O'} + \vec{r}_{O'/O} . \tag{3.1}$$

Again, the notation "O'/O" means O' relative to O. Note that $\vec{r}_{O'/O} = -\vec{r}_{O/O'}$. This gives us the equation for relative displacement.

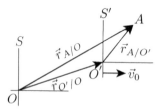

Fig. 3.1

In the case where the relative motion of the reference frame is much less than the speed of light (i.e. in the classical non-Einsteinian regime), the time interval in the S and S' frames is the same:

$$\Delta t = \Delta t' . \tag{3.2}$$

The invariance in time interval allows us to differentiate both sides of Eq. (3.1) with respect to time, leading to

$$\vec{v}_{A/O} = \vec{v}_{A/O'} + \vec{v}_{O'/O} . \tag{3.3}$$

This gives the equation for the relative velocity, where we again note that $\vec{v}_{O'/O} = -\vec{v}_{O/O'}$.

By the same argument above, we differentiate both sides of Eq. (3.3) with respect to time. We obtain

$$\vec{a}_{A/O} = \vec{a}_{A/O'} + \vec{a}_{O'/O} , \tag{3.4}$$

which is the equation of relative acceleration. Note that $\vec{a}_{O'/O} = -\vec{a}_{O/O'}$.

3.4   Examples of Relative Motion

In this section, we illustrate the applicability of relative motion as an approach as well as an alternative method, in the solution of kinematic problems. The usefulness of relative motion in this respect comes from viewing the problem through the perspective of different reference frames, which not only helps to simplify the formulation of the problem, but also clarifies the physical condition of the problem, contributing to its appropriate solution. These aspects of relative motion will be demonstrated through three kinds of relative motion: *relative displacement*, *relative velocity*, and *relative acceleration*.

3.4.1   *Relative Displacement*

In section 2.7.2, we have obtained the range R of a projectile over an inclined plane via a geometric approach with vector analysis. Here, we present an alternative method to evaluate R by using the idea of relative displacement as well as the equation of kinematics of translational motion. To begin, we define two frames of reference. One reference frame S is attached to the ground with its origin O set at the launching point of the projectile. This origin is also located at the edge of the inclined plane as shown in Fig. 3.2. The second reference frame S' has its axes parallel to S but moves with a constant velocity of $v_i \cos(\theta + \beta)$ to the right relative to O in S. At time $t = 0$, both the origin of S and S' coincides, and at this instant, the projectile is launched. Since the horizontal velocity of the projectile is $v_i \cos(\theta + \beta)$ to the right relative to S, its horizontal motion coincides with S'. Thus, the projectile is perceived to have zero horizontal velocity relative to S'. According to S', the projectile was launched vertically at its origin O' with an initial velocity of $v_i \sin(\theta + \beta)$. S' will describe the displacement of the ball from O' with time t as follows:

$$s = [v_i \sin(\theta + \beta)]\, t - \frac{1}{2}gt^2\,. \tag{3.5}$$

Moreover, by examining the geometry of the problem, we observe that the displacement of the surface of the inclined plane from O' depends on time t in the following manner:

$$s_h = r \tan \beta = v_i t \cos(\theta + \beta) \tan \beta\,, \tag{3.6}$$

where $r = v_i t \cos(\theta + \beta)$. When the projectile hits the inclined plane at time $t = t_R$, $s = s_h|_{t=t_R}$. Substituting the above equations according to

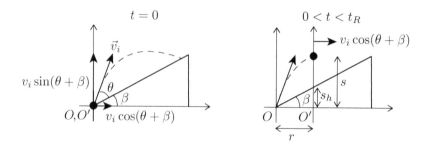

Fig. 3.2

these conditions, we have

$$v_i t_R \sin(\theta + \beta) - \frac{1}{2} g t_R^2 = v_i t_R \cos(\theta + \beta) \tan \beta \,.$$

Solving this equation for t_R, we obtain

$$t_R = \frac{2v_i}{g} \left[\sin(\theta + \beta) - \cos(\theta + \beta) \tan \beta \right] \,,$$

$$= \frac{2v_i}{g \cos \beta} \left[\sin \theta \cos^2 \beta + \sin \theta \sin^2 \beta \right] \,,$$

which gives

$$t_R = \frac{2v_i \sin \theta}{g \cos \beta} \,. \tag{3.7}$$

Next, we observe that the relative displacement equation of the problem: $\vec{r}_{B/O} = \vec{r}_{O'/O} + \vec{r}_{B/O'}$ is represented by a right-angled triangle as shown in Fig. 3.3, such that the range R can be easily deduced as follows. First, note that

$$\sin \beta = \frac{|\vec{r}_{B/O'}|_{t=t_R}}{|\vec{r}_{B/O}|_{t=t_R}} = \frac{s_h|_{t=t_R}}{R} \,.$$

Rearranging terms, we have

$$R = \frac{s_h|_{t=t_R}}{\sin \beta} \,,$$

$$= \frac{v_i t_R \cos(\theta + \beta) \tan \beta}{\sin \beta} \,,$$

$$= \frac{v_i \cos(\theta + \beta) \tan \beta}{\sin \beta} \left(\frac{2v_i \sin \theta}{g \cos \beta} \right) \,,$$

$$= \left(\frac{2v_i^2}{g} \right) \cos(\theta + \beta) \left(\frac{\sin \beta}{\cos \beta} \right) \left(\frac{\sin \theta}{\cos \beta} \right) \left(\frac{1}{\sin \beta} \right) \,,$$

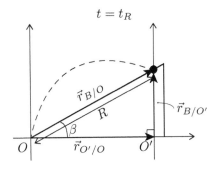

Fig. 3.3

which leads to

$$R = \frac{2v_i^2 \sin\theta \cos(\theta + \beta)}{g\cos^2\beta}. \qquad (3.8)$$

Note that this is the same R as that found in Eq. (2.32).

3.4.2 Relative Velocity

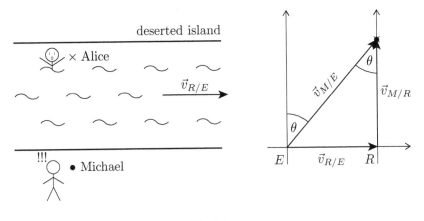

Fig. 3.4

In this example, we have the case of Michael ($\bullet$) saving Alice ($\times$) using the rule of relative velocity under two scenarios. In both scenarios, we have a flowing river. The nearer river bank is where Michael is and the remote river bank belongs to a deserted island. Assuming that the velocity of the river relative to the river bank (or Earth) $\vec{v}_{R/E}$ is to the right with magnitude

$|\vec{v}_{R/E}| = 5\,\text{km/h}$, and Michael can swim with a speed of $|\vec{v}_{M/R}| = 10\,\text{km/h}$ relative to the river, how can he rescue Alice whom he sees drowning directly ahead of him? Because the speed of Alice relative to the river is zero, Michael should swim directly across the river towards Alice in order to save her. The relative motion of this situation is depicted in Fig. 3.4. By means of relative velocity:

$$\vec{v}_{M/E} = \vec{v}_{M/R} + \vec{v}_{R/E}\,, \qquad (3.9)$$

an observer on the nearer river bank would have observed Michael to have traversed at a speed of

$$|\vec{v}_{M/E}| = \sqrt{10^2 + 5^2} = 11.2\,\text{km/h}\,.$$

His direction of travel to the observer can be worked out as follows:

$$\tan\theta = \frac{5}{10} \implies \theta = 26.6°\,,$$

indicating a direction of 26.6° east-of-north.

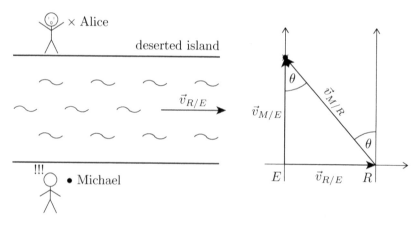

Fig. 3.5

In the second scenario, Alice is trapped in the deserted island and she is crying for help at the river bank directly across from Michael (see Fig. 3.5). In order to save Alice at where she is, Michael should not swim directly across the river. Using the relative velocity relation given by Eq. (3.9), Michael constructs a vector triangle as shown in Fig. 3.5. This vector triangle advises Michael to head in a direction θ degree west-of-north. Performing the calculation based on this vector triangle, we have

$$\sin\theta = \frac{|\vec{v}_{R/E}|}{|\vec{v}_{M/R}|} = \frac{5}{10} \implies \theta = 30°\,.$$

Thus, Michael should swim in the direction of 30° west-of-north to save Alice.

3.4.3   Relative Acceleration

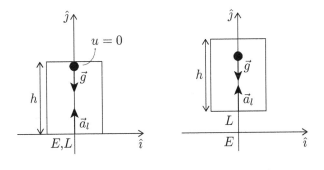

Fig. 3.6

Finally, we consider the problem of a ball falling from the ceiling of a lift as shown in Fig. 3.6. At time $t = 0$, the ball is dropped from rest with acceleration $\vec{a}_{B/E} = \vec{g} = -g\hat{j}$ relative to Earth. At the same time, the lift starts from rest and accelerates upward with $\vec{a}_{L/E} = \vec{a}_l = a_l\hat{j}$ relative to Earth. If we were to take reference from Earth, we can determine the time t at which the ball hits the floor of the lift by means of the kinematic equation:

$$s = ut + \frac{1}{2}at^2\,.$$

Let d_1 and d_2 be the distance traveled by the ball and the lift respectively at the instant when the ball hits the floor of the lift. Noting that $|\vec{a}_l| = a_l$ and $|\vec{g}| = g$, we expect

$$d_1 = \frac{1}{2}gt^2\,,$$
$$d_2 = \frac{1}{2}a_lt^2\,.$$

The sum of these two distances is the height h of the lift:

$$h = d_1 + d_2 = \frac{1}{2}\left(g + a_l\right)t^2\,,$$

such that

$$t = \sqrt{\frac{2h}{g + a_l}}\,.$$

However, we can also yield the time t by taking reference from the lift instead of the Earth. To do this, we employ the following relation from relative acceleration:

$$\vec{a}_{B/E} = \vec{a}_{B/L} + \vec{a}_{L/E}\,, \tag{3.10}$$

to obtain the acceleration of the ball relative to the lift,

$$\begin{aligned}
\vec{a}_{B/L} &= \vec{a}_{B/E} - \vec{a}_{L/E}\,, \\
&= \vec{g} - \vec{a}_l\,, \\
&= -g\hat{j} - a_l\hat{j}\,, \\
&= -\left(g + a_l\right)\hat{j}\,.
\end{aligned}$$

Thus, the ball accelerates downward with

$$|\vec{a}_{B/L}| = g + a_l$$

relative to the lift. Using the kinematic equation

$$s = ut + \frac{1}{2}at^2$$

again, with $u = 0$ and $a = |\vec{a}_{B/L}|$, we have

$$h = \frac{1}{2}|\vec{a}_{B/L}|t^2 = \frac{1}{2}\left(g + a_l\right)t^2\,.$$

Solving for t, we obtain

$$t = \sqrt{\frac{2h}{g + a_l}}\,, \tag{3.11}$$

which is the same as the earlier result taking reference from Earth.

As a final note, the accelerating reference frame of the lift considered here is in fact a non-inertial reference frame[1]. Conversely, the analysis earlier with the reference frame of the Earth corresponds to that performed on an inertial reference frame. It is interesting that by applying two different

---

[1]Note that both the non-inertial reference frame and the inertial reference frame will be introduced in the next section and Chapter 4.

approaches: one employing the inertial reference frame and the other using the non-inertial reference frame, we arrive at the same answer through taking different perspectives on the same problem.

3.5 Galilean Relativity

The first person to state the principle of relativity is not Albert Einstein. It was first stated by Galileo Galilei as follows: *The laws of physics are the same in any inertial frame of reference.*

To understand this statement, we need to know what is an inertial frame of reference[2]. The inertial frame of reference can be defined from the perspective of the reference frame we discussed in section 3.2. If the reference frame S in section 3.2 is an inertial reference frame, and $\vec{v}_{O'/O}$ is constant, then S' is also an inertial reference frame. Under this condition, $\vec{a}_{O'/O} = 0$.

More specifically, let $\vec{v}_{O'/O} = \vec{v}_0$, with $\vec{v}_0$ being a constant vector. If we were to define $t = 0$ to be the instant when the origins of the reference frame S and S' coincide, the event A can be described as follows:

$$\vec{r}_{A/O} = \vec{r}_{A/O'} + \vec{v}_0 t, \tag{3.12}$$

$$t_{A/O} = t_{A/O'} = t. \tag{3.13}$$

Equations (3.12) and (3.13) are known as the *Galilean coordinate transformation.*

We can employ the Galilean coordinate transformation to measure time interval and space interval. Equation (3.13) tells us that the time interval between the occurrence of two given events, say A and B, is the same for each observer:

$$t_{A/O} - t_{B/O} = t_{A/O'} - t_{B/O'}. \tag{3.14}$$

On the other hand, space interval is the magnitude of the displacement between two events which occur at the same time but different points in space. According to Eqs. (3.12) and (3.13), the space interval of events A and B relative to O and O' is the same:

$$|\vec{r}_{A/O} - \vec{r}_{B/O}| = |\vec{r}_{A/O'} - \vec{r}_{B/O'}|. \tag{3.15}$$

---

[2]The physical concept of the inertial frame of reference will be more accurately defined when we reach the topic of Newton's first law.

These results illustrate the absoluteness of the space and time interval of non-Einsteinian physics.

The fact that S and S' move in a constant relative velocity $\vec{v}_0$ implies that

$$\vec{v}_{A/O} = \vec{v}_{A/O'} + \vec{v}_0 \tag{3.16}$$

according to Eq. (3.3). This equation is known as the *Galilean velocity transformation*. It shows that the observers in the two frames measure different velocities for the same event.

Finally, we apply Eq. (3.4) and since $\vec{v}_{O'/O} = \vec{v}_0$ is a constant relative velocity, $\vec{a}_{O'/O} = 0$. Thus

$$\vec{a}_{A/O} = \vec{a}_{A/O'}\,, \tag{3.17}$$

i.e., the two observers measure the same acceleration for the same event A if they move at a constant velocity relative to each other. Let us now consider event A to be the motion of a particle of mass[3] m. If we were to assume that the mass of this particle is not affected by the motion of the reference frame, then $m_{A/O} = m_{A/O'} = m$, where $m_{A/O}$ and $m_{A/O'}$ is the mass of particle A in the inertial reference frame S and S' respectively. By combining this assumption with Eq. (3.17), we obtain the following relation

$$m_{A/O}\vec{a}_{A/O} = m_{A/O'}\vec{a}_{A/O'}\,. \tag{3.18}$$

Placing this result in the context of Galilean principle of relativity, the following conjecture for a law of force $\vec{F}$ could be arrived:

$$\vec{F} = m\vec{a}\,. \tag{3.19}$$

This law of force has the characteristic of remaining invariant (i.e. the same) between the different inertial reference frames. Could this feature be the reason that Newton formulated his laws of mechanics the way it is?

---

[3]The definition of mass will be given in the next chapter in the section on Newton's second law.

Chapter 4

Newton's Laws

4.1 Introduction

We have studied kinematics and the subject has provided a physical quantification of *motion*, and vectors have been shown to serve a comprehensive mathematical framework for this purpose. At this point, it is appropriate to ask how motion arises. The answer is a new physical quantity that we termed *force*. Force is the *cause*, while acceleration is its consequential *effect* which accounts for the origin of motion.

4.2 Real Forces

It is known that there are four fundamental forces or interactions in nature[1]. These are the:

(1) Gravitational Force
(2) Electromagnetic Force
(3) Strong Force
(4) Weak Force

We term them as the *real* forces in order to distinguish from the *fictitious* force that we will introduce later in the chapter. All other forces such as friction, intermolecular forces, van der Waal forces are derived from these fundamental and real forces. The gravitational and electromagnetic forces are long range forces of infinite extent encountered in our daily life. The strong and weak forces are nuclear forces which arise at subatomic distances. In classical physics, it is possible to describe the gravitational and

---

[1] The electromagnetic force and the weak force have been unified into one electroweak force at high energies.

electromagnetic forces by Newton's laws. It is, however, not possible to express the strong and weak forces by Newton's laws. These forces can only be formulated through quantum field theory. In fact, the electromagnetic force is also intrinsically quantum mechanical in nature while the gravitational force can be more precisely defined through the theory of general relativity. These non-Newtonian characteristics are a more accurate description of the forces which lie in the realm of modern physics.

4.3   Newton's First Law

Newton's first law or the law of inertia states that

In the absence of net (or resultant) external forces, when viewed from an inertial reference frame, an object at rest remains at rest and an object in motion continues in motion with a constant velocity (that is, at a constant speed in a straight line).

There are two aspects to the first law. One aspect is to clearly define the concept of *inertia*. The first law describes the natural tendency of an object not to change its state of motion. This resistance to change then allows the definition of the fundamental physical quantity that we call *mass*. Here, the resistance to change refers to a change with respect to the state of uniform velocity: the state of rest or the state of nonzero constant velocity are both states of uniform velocity. Specifically, the more difficult for an object to change from this uniform state, the greater is the mass of the object.

You may have noticed the seeming equivalence between the state of rest and the state of constant velocity in Newton's first law. This is the second aspect which serves to define the reference frame for the observation of the motion of the object *in the absence of external force*. Let us assume that there is no external force acting on the object, and that an observer O is in a reference frame S which observes this object to be in a state of rest. Then, reference frame S serves to define an inertial reference frame. Moreover, any reference frame moving at constant velocity relative to this frame is also an inertial reference frame according to our discussion in section 3.5. Using any of the latter set of reference frames to observe the same object, one would observe the object to be traversing in constant velocity. Thus, from the perspective of the inertial reference frames, one perceives the context of the equivalence between the state of rest and the state of constant velocity in the first law. In this way, Newton's first law defines the set of inertial reference frames, and you will see later that Newton's laws are in fact invariant within these set of inertial reference frames with respect to the real forces.

4.4 Newton's Second Law

Newton's second law states that

When viewed from an inertial reference frame, the acceleration of an object is directly proportional to the net (or resultant) real forces acting on it and inversely proportional to its mass:

$$\sum \vec{F} = m\vec{a}. \tag{4.1}$$

A direct meaning of the second law is that force is the cause of a change in motion (i.e. the effect). Moreover, Eq. (4.1) shows that force is a vector since acceleration is a vector. It is important to note that the forces we are referring to here are the real forces. $\sum \vec{F}$ is the vector sum of these real forces that act on the same object. We term $\sum \vec{F}$ the *net* force or the *resultant* force. Figure 4.1(a) illustrates a distribution of these forces that act on different part of an object. In our treatment of the object as a particle, we have brought all the forces to a particular point in the object[2] as shown in Fig. 4.1(b). We then perform vector addition on all these forces to obtain a resultant. We can in fact replace the group of forces in Fig. 4.1(b) with this resultant force without affecting the object's motion. This resultant force then dictates the translational acceleration of the object.

Equation (4.1) also gives a more quantitative definition of the inertia mass m introduced in Newton's first law. Note that the SI unit of mass is the kilogramme (kg). As a result, the SI unit of force is kilogramme meter per second squared (kg m/s^2). The physical meaning of the inertia mass here is different from the gravitational mass m' in Newton's law of universal gravitation $F = GMm'/r^2$. While the inertia mass measures the resistance to a change in the state of motion of an object, gravitational mass corresponds to the strength of the gravitational field emitted by the

---

[2]This point is in fact the center of mass of the object, which will be shown when we deal with extended object in Chapter 6. Note that displacing all the forces to the COM does have physical effects as indicated in Chapter 1, and will be clearly illustrated in Chapter 7 that the effects are rotational. In the current context, these rotational effects are ignored as we are focusing on translational motion.

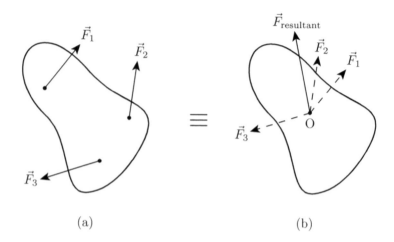

Fig. 4.1

object concerned. In consequence, these two masses are very different conceptually. Nonetheless, the theory of general relativity is able to prove that they are in fact equivalent.

An important remark on Eq. (4.1) is that it is valid for real forces in inertial reference frames. For Eq. (4.1) to continue to be applicable in the non-inertial reference frames, fictitious forces need to be included. This will be discussed in detail in section 4.9.

4.5   Newton's Third Law

Newton's third law states that

If two objects interact, the force $\vec{F}_{2\to1}$ exerted on object 1 by object 2 is equal in magnitude and opposite in direction to the force $\vec{F}_{1\to2}$ exerted on object 2 by object 1:

$$\vec{F}_{2\to1} = -\vec{F}_{1\to2}\,. \tag{4.2}$$

The important point to note is that Newton's third law relates to the interaction between two bodies (see Fig. 4.2). The pair of forces involved is known as the *action* and *reaction* pair.

Figure 4.3 shows an example on how the action and reaction pair (or "third-law pair") appears for a TV placed on a table, and the table rests on

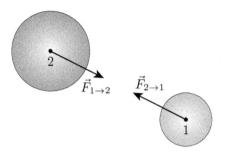

Fig. 4.2

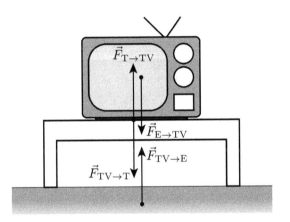

Fig. 4.3

solid ground. The TV is acted on by the gravitational force from the Earth $\vec{F}_{E \to TV}$. By Newton's third law, the Earth shall experience an equal and opposite force $\vec{F}_{TV \to E}$ from the TV. In addition, the TV is also acted on by the normal force from the table $\vec{F}_{T \to TV}$. Newton's third law then dictates the action of an equal and opposite force $\vec{F}_{TV \to T}$ on the table. These forces are clearly indicated in Fig. 4.3. Similarly, the table is also exerted by forces from the Earth and the ground. And in turn, the Earth and ground experience the equal and opposite reaction forces from the table[3]. These forces are all real, with the normal forces being electromagnetic in origin.

---

[3]To avoid Fig. 4.3 being too cluttered, we have not drawn the action and reaction pairs between the table and the Earth, and the table and the ground.

Notably, Newton's third law only applies to real forces. In particular, the action and reaction forces are both real. They act on different objects and must be of the same type. A single isolated real force cannot exist.

4.6 Friction and Resistive Forces

When an object is moving on a surface or through a viscous medium, it will encounter resistance to its motion. This results from the interactions between the object and its environment. Typically, the interaction comes in two forms. If the object moves in contact with a surface, its motion is impeded by a force known as *friction*. On the other hand, if the object moves through a viscous medium, it is subjected to *resistive forces*. Note that we shall go into greater detail on friction, and less on resistive forces in this book.

Friction arises as a result of the molecular interactions between two surfaces. It is a many-body interaction and it is basically electromagnetic in origin. Physically, the surfaces of the two materials in contact are not smooth microscopically. They touch each other via many microscopic bumps. As the two surfaces slide over each other, the bumps on one surface block the motion of the bumps on the opposing surface through electromagnetic interactions between the atoms and molecules in the bumps. In addition, the molecules within the two surfaces can form chemical bonds between them which causes the surfaces to stick together. All these interactions prevent motion or lead to a resistance to motion which collectively form an effective force which we call friction.

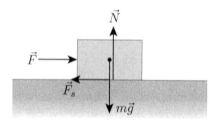

Fig. 4.4

It is important to note that there are two types of friction: *static friction* and *kinetic friction*. Static friction acts when there is no relative motion between the two surfaces in contact. On the other hand, kinetic (or sliding)

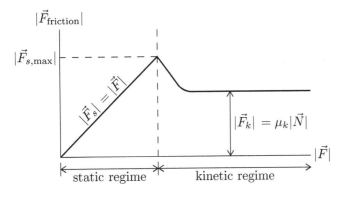

Fig. 4.5

friction happens when the two surfaces slide over each other. Let us take an example to develop a deeper understanding on these two forms of friction. Imagine an object of mass m resting on the ground. For this to happen, the normal force $\vec{N}$ from the ground acting on the object must be equal and opposite to the gravitational force $m\vec{g}$ (see Fig. 4.4) so that the object does not move. Now, we start to apply a horizontal force $\vec{F}$ rightward with increasing magnitude on this object. Initially, the object remains motionless. It is at this phase that static friction $\vec{F}_s$ is acting, such that $\vec{F}_s$ is equal in magnitude but opposite in direction to $\vec{F}$. As the magnitude of $\vec{F}$ increases (decreases), the magnitude of $\vec{F}_s$ also increases (decreases) so that the net force acting on the object vanishes:

$$\vec{F} + \vec{F}_s = 0 \,. \tag{4.3}$$

The static friction has successfully prevented motion of the object at this stage when the applied force is not sufficiently large. However, when the magnitude of $|\vec{F}|$ reaches the limit of the maximum static frictional force which has magnitude

$$|\vec{F}_{s,\max}| = \mu_s |\vec{N}| \,, \tag{4.4}$$

where μ_s is the coefficient of static friction, the object is now perched at the verge of sliding. At this point, $\vec{F}_{s,\max}$ just barely cancels $\vec{F}$. When $|\vec{F}| > \mu_s |\vec{N}|$, static friction can no longer resist motion and the object starts to slide. The interaction between the surface of the object and the ground is now in the regime of kinetic friction. Figure 4.5 displays the details of such a transition from static to kinetic friction. While chemical bonds or the so-called "spot welds" are the reason of static friction as the surfaces become bound together, the collision of the microscopic bumps over each other when the two surfaces slide account for kinetic friction. Thus, the kinetic friction $\vec{F}_k$ acts when the object is in motion. Although the magnitude of kinetic friction does vary with speed, it is only weakly so. Hence, we can assume its magnitude $|\vec{F}_k|$ to be constant, i.e. independent on velocity. Specifically, we have

$$|\vec{F}_k| = \mu_k |\vec{N}|\,, \tag{4.5}$$

where μ_k is the coefficient of kinetic friction, and always $\mu_k < \mu_s$. Like $\vec{F}_s$, $\vec{F}_k$ is tangential to the surface where friction occurs. Both $\vec{F}_s$ and $\vec{F}_k$ act not only in a direction that opposes motion (see Fig. 4.4 for $\vec{F}_s$), they can also act in a direction that creates motion (see Fig. 4.6 for $\vec{F}_s$)[4]

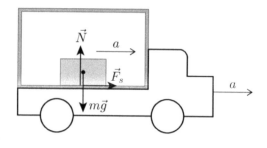

Fig. 4.6

One important property of friction is that it is independent of the area of the contact surface on which it acts. This can be understood as follows. Assume an object resting (or moving) on the floor. If its surface area in contact with the floor is reduced by a factor of $1/k$, its weight per unit area pressing down on the floor will increase by a factor of k. The implication is that the frictional force per unit area will also increase k times. To obtain the frictional force, the frictional force per unit area is to be multiplied by its contact area. The consequence is that the k-times enhanced frictional force per unit area is exactly neutralized by the $1/k$-times reduced area, demonstrating that objects of equal mass but with different contact area all have the same frictional force, i.e. frictional force is independent of its contact area.

---

[4]An example where $\vec{F}_k$ acts in the direction of motion is exhibited in Chapter 7 section 7.3.6.4 where a rotating sphere is placed onto a ground with friction (see Fig. 7.15). Here, $\vec{F}_k$ generates acceleration in the direction of translational motion while creating a torque that opposes rotational motion.

A final comment on the frictional force is that it has a very interesting physical characteristic that is different from the gravitational force or the elastic spring force (which is electromagnetic in origin) discussed in this book. In problems or exercises encountered here or elsewhere, the latter forces typically act between single bodies. But friction is a many-body effective force and to decipher its physical details requires the science of thermodynamics or the tools of statistical mechanics. Essentially, friction embodies the many-body features through its expression of the notion of *irreversibility* in the effects that it generates. It would be good to clarify this with an example.

We know that the kinematic effect of the gravitational force is reversible in time. Let us see this by taking a video of a body which is initially at rest dropping from height h. We observe the body picks up speed with a constant acceleration $\vec{g}$ and attains a velocity $\vec{v}$ just before it touches the ground. At this point, we stop the video. By playing the video backward in time, we now observe that the body moves upward against gravity with initial velocity $-\vec{v}$ and decelerate as $\vec{g}$. It finally comes to a stop at height h. We observe that the reverse motion is physically tenable since it can be easily demonstrated through an experiment. Now, let us take a video of a block sliding on a floor with initial velocity $\vec{u}$. The block decelerates due to kinetic friction and eventually comes to rest. If we were to play the video backward in time, we observe the block starts from rest spontaneously on its own accord and miraculously accelerates to a speed $-\vec{u}$ subsequently. This phenomenon is never observed in nature. Furthermore, it is totally unphysical to attribute the cause of the acceleration of the block to friction. But we can clearly account for what we see in the first video to that of gravitational force. Here thus lies the fundamental difference between the *single-body*[5] gravitational force and the *many-body* frictional force discussed in the two examples. While the former gives rise to reversible kinematic effects, the latter possesses kinematic effects that are irreversible. In fact, the irreversible effects from friction are intrinsically related to the second law of thermodynamics.

Finally, there is another class of force that dampens motion and is termed the resistive force. It is also a type of many-body force like friction. Unlike frictional force, however, resistive force arises from the interaction between the object and the medium through which it moves. The medium

---

[5]The main point here is that there is no effect of statistical randomization and entropy change for single-body forces, while these effects occur naturally for the many-body forces.

can either be a liquid or a gas. Examples of resistive forces are air resistance or the drag force faced by an object moving through a liquid. Let us consider an object moving through the medium, with the medium exerting a resistive force $\vec{R}$ on it. $\vec{R}$ is an effective force that acts in a direction opposite to the motion of the object relative to the medium. Its magnitude depends on the medium and the speed of the object. Specifically, the magnitude of $\vec{R}$ increases with the speed of the object. There are two different formulations of the resistive forces depending on the size of the object. For many microscopic objects such as dust particles, pollen grains, bacteria, viruses, colloidal particles, bubbles, macromolecules etc, the magnitude of the resistive force is proportional to the speed of the object. On the other hand, objects of macroscopic sizes such as cars, aeroplanes, soccer balls, cyclists etc, encounter a resistive force with magnitude that depends on the squared speed of the object. In particular, the effects of turbulence is pertinent to the drag force experienced by these macroscopic objects. We can also have $|\vec{R}| \propto v$ for low v, and $\propto v^2$ for high v, for the *same* object. Furthermore, the dynamics of the object subjected to resistive force is different from that due to frictional force in one key aspect. While the resultant acceleration of the object under friction is typically constant, the acceleration of an object acted on by a resistive force is generally time dependent. In the steady state, the object under a resistive force is found to attain a terminal velocity as its acceleration approaches zero.

4.7   Centripetal Force

Recall our earlier study on a type of acceleration known as the centripetal acceleration. Such an acceleration does not change the speed of an object but changes its direction, leading to uniform circular motion. Again, it is the real force that causes the centripetal acceleration. For example, the real force that causes the centripetal acceleration of the conical pendulum in Fig. 4.7(a) is the tension of the string which is electromagnetic in origin. More specifically, the centripetal force is $T \sin \theta$ which lies in the horizontal plane and directs towards the center of the circle. Similarly, the centripetal force on the car driving through a ramp as illustrated in Fig. 4.7(b) is the component $n \sin \theta$ of the normal force which is also intrinsically an electromagnetic force. Finally, Fig. 4.7(c) shows that static friction can also serve as centripetal force that enables the car to turn in a circle.

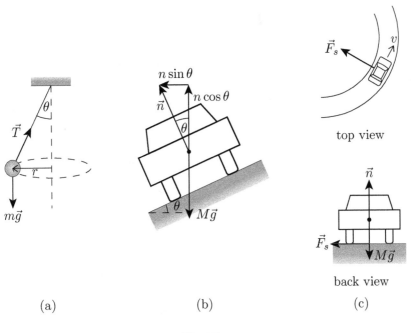

top view

back view

(a)   (b)   (c)

Fig. 4.7

4.8   An Example on the Application of Centripetal Force and Static Friction

In this section, we show an interesting real-world application of centripetal force with static friction that comes in the form of an amusement park ride. The ride consists of a large vertical cylinder that rotates about its axis sufficiently fast so that any person inside it remains held up against the wall even when the floor drops away (see Fig. 4.8). In order for the ride to operate the way it is, three real forces are observed to act on each of the park goers. The first is the gravitational force Mg of the Earth pulling a person of mass M vertically downward. The second is the normal force N from the cylinder pushing on the back of the person, with the force directing radially inward towards the central axis of the cylinder. The third is a static friction force F_s of the cylinder on the clothing of the person since there is no relative motion between the person and the wall. Because the person tends to slide down the wall, this force acts vertically upward. We shall denote the coefficient of static friction between the clothing of the person and the wall as μ_s, and the radius of the cylinder as R. Note that the second and third forces are intrinsically electromagnetic in nature.

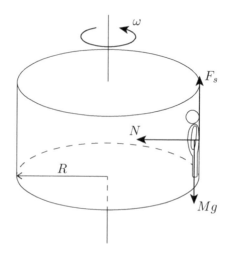

Fig. 4.8

The physical condition of the ride on the person is described by the following set of equations according to Newton's second law:

$$F_s = Mg \,, \tag{4.6}$$

$$N = MR\omega^2 \,. \tag{4.7}$$

The first equation indicates that there is no net vertical force acting on the person, ensuring that the person maintains a constant height above the sunken floor as the cylinder rotates with an angular velocity ω about the axis of the cylinder. On the other hand, Eq. (4.7) shows that the normal force N supplied the centripetal force of the person's uniform circular motion.

Let us now discuss some interesting subtleties that underlie the design of this amusement ride. For that purpose, we first perform an analysis based on $F_s \leq F_{s,max}$, where $F_{s,max} = \mu_s N$. Using Eq. (4.7), we obtain $F_s \leq \mu_s MR\omega^2$, which leads to

$$\omega \geq \sqrt{\frac{g}{\mu_s R}} \tag{4.8}$$

after employing Eq. (4.6). In other words,

$$\omega_{min} = \sqrt{\frac{g}{\mu_s R}} \,. \tag{4.9}$$

If we assume μ_s to be the same for all customers who take this ride, the angular velocity ω of the cylinder must be larger than ω_{min} to ensure that none of the customer slides down the wall. Notice that this condition is true for any mass M of the customers, which is the key criterion that enables the operationalization of this amusement ride. In other words, if $\omega < \omega_{min}$, all the customers will slide down the wall regardless of their mass.

Now, what if one of the customers were to wear a clothing made of satin, which has a smaller μ_s relative to the other clothing materials? Based on Eq. (4.8), we first deduce that

$$\mu_s \geq \frac{g}{R\omega^2} \tag{4.10}$$

for all customers to remain fixed to the wall. If only the satin-clothed customer has $\mu_s < g/R\omega^2$, we expect this customer to be the only one sliding down the wall. This could have disastrous consequences. For the safety of all customers, the amusement park may require all customers to wear a vest to ensure the condition given by Eq. (4.10) is satisfied at all time as part of this ride's protocol.

4.9 Non-inertial Reference Frame and the Fictitious Force

We have discussed Newton's laws and its application with respect to inertial reference frames in sections 3.5 and 4.4. The question we now have is: can we apply Newton's laws when the frame is non-inertial?

To address this question, we consider an observer O in an inertial reference frame S observing the motion of a particle A of mass m. Assuming a real force $\vec{F}_R$ acts on this particle, and because S is an inertial reference frame, Newton's second law is applicable and we have

$$\vec{F}_R = m\vec{a}_{A/O} . \tag{4.11}$$

Now, let O' be another observer making observations on the same particle A with respect to his reference frame S'. But S' is a non-inertial reference frame because it accelerates at an acceleration $\vec{a}_{O'/O}$ with respect to S. In fact, *any reference frame that accelerates with respect to an inertial reference frame defines a non-inertial reference frame*. Then, by means of Eq. (3.4), we obtain the following relation between the accelerations:

$$\vec{a}_{A/O} - \vec{a}_{O'/O} = \vec{a}_{A/O'} . \tag{4.12}$$

Multiplying both sides of the equation by m, we have

$$m\vec{a}_{A/O} - m\vec{a}_{O'/O} = m\vec{a}_{A/O'} . \tag{4.13}$$

Inserting Eq. (4.11) into this equation, we get

$$\vec{F}_R - m\vec{a}_{O'/O} = m\vec{a}_{A/O'} . \tag{4.14}$$

Notice that we have made the product of m and the acceleration of A with respect to O' the subject on the right-hand side of the equation. This formulation allows us to clearly see that Newton's second law is inapplicable in a non-inertial reference frame due to the additional term $-m\vec{a}_{O'/O}$ on the left-hand side of the equation. This term arises purely from the relative acceleration between frame S and S' and has nothing to do with the real forces. However, we can make Eq. (4.14) follow Newton's second law if we were to define a new type of "force"

$$\vec{F}_f = -m\vec{a}_{O'/O} . \tag{4.15}$$

Note that this is artificial but nonetheless, by creating this new force and including $\vec{F}_f$ into Eq. (4.14), we obtain

$$\vec{F}_R + \vec{F}_f = m\vec{a}_{A/O'} . \tag{4.16}$$

Thus, in order to apply Newton's second law in the non-inertial reference frame, we must include additional "forces" that result from the relative acceleration of the non-inertial frame with respect to the inertial frame. Such force is termed the *fictitious force* or the *pseudo-force*. By considering the net force as a vector sum of real and fictitious forces, Eq. (4.16) has generalized the application of Newton's second law to the non-inertial reference frame.

The derivation given above is for relative acceleration motion that purely translates. The same result is obtained when the relative acceleration purely rotates. Equation (4.16) is again obtained with $\vec{F}_f$ now consisting of two fictitious forces: the centrifugal force and the Coriolis force. The details of the derivation are given in Appendix A. In general, when the relative acceleration both translates and rotates, Eq. (4.16) is still true with $\vec{F}_f$ now being a vector sum of the translational part (Eq. (4.15)) and the rotational part (Eqs. (A1) and (A2)).

It is important to reiterate here that a fictitious force is not a real force. Hence, it does not originate from the gravitational force[6], electromagnetic force, the strong or the weak force. A fictitious force does not have an action and reaction pair like a real force, and they arise purely from the relative acceleration of the reference frame.

---

[6]Note that this comment is taken in the context of Newtonian physics. In the theory of general relativity, the gravitational force is in fact viewed as a fictitious force.

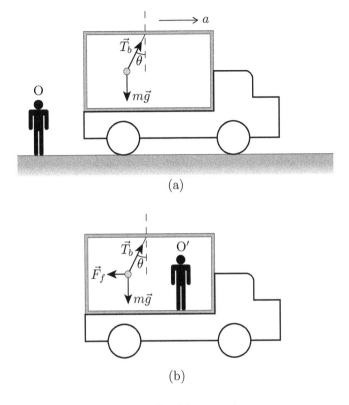

(a)

(b)

Fig. 4.9

It is instructive now to consider a few examples. First, let us consider an accelerating car with a bob and a string attached to it as shown in Fig. 4.9(a). To an observer O on the ground (which is an inertial reference frame) observing the bob, he can account for the fact that the string attached to the bob has to incline at an angle θ to the vertical. The car is accelerating to the right with an acceleration $\vec{a}$. The bob is observed to have the same acceleration $\vec{a}$ relative to observer O. It is obvious to O that the real force that causes the acceleration of the bob is the tension $\vec{T}_b$ in the string. More precisely, the tilting of the string allows the horizontal component of the tension to produce the acceleration:

$$|\vec{T}_b| \sin \theta = m|\vec{a}| \,, \tag{4.17}$$

where m is the mass of the bob. On the other hand, according to the non-inertial observer O' in the car, the string is also observed to tilt at the same angle θ to the vertical (see Fig. 4.9(b)). Furthermore, O' observes that the tension in the string is the same $\vec{T}_b$ as that measured by O. This can easily be checked by means of a spring balance. However, with just $\vec{T}_b$ and the weight of the bob $m\vec{g}$, he cannot account for the fact that the bob is stationary in his frame of reference. The only way he can explain this while preserving the validity of Newton's second law is to include a fictitious force $\vec{F}_f$ as follows:

$$\vec{T}_b + m\vec{g} + \vec{F}_f = 0. \tag{4.18}$$

Based on Eq. (4.15), $\vec{F}_f = -m\vec{a}$ and thus the fictitious force is a horizontal force pointing to the left. Putting this into Eq. (4.18), it is straightforward to evaluate and determine that

$$|\vec{T}_b| \sin\theta - m|\vec{a}| = 0, \tag{4.19}$$

which matches the result given by Eq. (4.17) for the inertial reference frame.

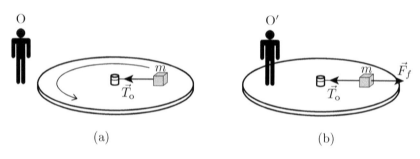

Fig. 4.10

Our second example involves an object of mass m rotating in a platform and is being pulled by a string (see Fig. 4.10(a)). According to the inertial observer O, the tension $\vec{T}_o$ in the string provides the centripetal force for the uniform circular motion of radius R of the object:

$$\vec{T}_o = -m\frac{v^2}{R}\hat{r}, \tag{4.20}$$

where v is the speed of the object and $\hat{r}$ is the unit radial basis vector of the polar coordinate system. To the non-inertial observer O', however, the object is stationary in his frame of reference as shown in Fig. 4.10(b). Nonetheless, O' does observe the presence of tension $\vec{T}_o$ acting through the string on the object, which can be easily verified by a spring balance. Thus, in order for O' to make sense of this result with respect to Newton's second law, he has to include a fictitious force (see Fig. 4.10(b)):

$$\vec{T}_o + \vec{F}_f = 0 \,. \tag{4.21}$$

Since the object is stationary in the non-inertial frame, i.e. $\vec{v}' = 0$, the Coriolis term is zero. Hence, $\vec{F}_f$ is just the centrifugal force $\left(mv^2/R\right)\hat{r}$ given by Eq. (A26). This leads to

$$\vec{T}_o + m\frac{v^2}{R}\hat{r} = 0 \,, \tag{4.22}$$

which is equivalent to Eq. (4.20).

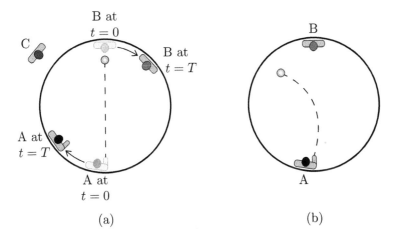

Fig. 4.11

In our last example, we give a simple illustration of the kinematic effect of the Coriolis force in a rotating reference frame. As shown in Fig. 4.11, two observers A and B are standing still in the rotating platform. At a moment in time, A throws a ball directly at B. According to an inertial observer C who is on the ground outside the rotating platform, the ball has a straight trajectory (see Fig. 4.11(a)). But according to A and B, the ball executes a curved trajectory. The curved path of the ball arises purely from the rotation of the platform, which can be easily discerned from Fig. 4.11(b). As the ball is flying directly to B, the rotation of the frame brings B away from the ball causing the ball to head in a direction to the right of B according to both non-inertial observers. The overall motion is thus a curved path according to these observers and to account for it, they have to introduce the Coriolis force (see Eq. (A26)).

In summary, fictitious forces do lead to observable kinematic effects in non-inertial reference frames. For example, the translational fictitious force acts to cause an object to accelerate in a direction opposite to the acceleration of the frame. Its effect is typically experienced when a car suddenly accelerates causing loose objects on the dashboard to fly (accelerate) towards the driver. Similarly, as a car executes a turn, the driver tends to feel an outward push by the centrifugal force acting in a direction that is away from the center of the turn. Finally, the Coriolis force causes the deflection of winds which directly impacts the occurrence of hurricane and storms in the Earth's climatic system. In conclusion, while the fictitious forces are fake forces, they do have real effects.

4.10   Examples on the Application of Non-Inertial Reference Frame

4.10.1   *Object Sliding Up an Accelerating Wedge*

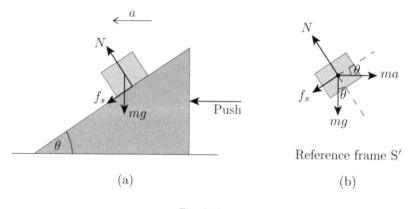

(a)   (b)

Fig. 4.12

A block of mass m is placed on the incline of a wedge whose angle of inclination with the horizontal is θ (see Fig. 4.12(a)). Note that there is friction between the surface of the block and the wedge with the coefficient of static friction being μ_s. A force drives the wedge such that it moves with a constant acceleration a to the left. Beyond a certain maximum acceleration a_{max} of the wedge, the object will slide up the incline. Our goal is to determine a_{max}. Let us take the perspective of a non-inertial reference frame S' that is attached to the wedge. S' thus accelerates to the left with acceleration a relative to an inertial reference frame that is attached to the ground. To proceed with our analysis and to apply Newton's second law, we include a fictitious force ma that is opposite in direction to the acceleration a based on section 4.9 (see Fig. 4.12(b)). With this, we formulate two equations of motion. The first equation considers the components of the forces that are normal to the inclined plane:

$$N = ma\sin\theta + mg\cos\theta, \tag{4.23}$$

while the second equation treats the components that are parallel to the inclined plane:

$$ma\cos\theta = f_s + mg\sin\theta. \tag{4.24}$$

Notice that in our formulation, we have assumed the static friction f_s points down the inclined plane. From Eq. (4.24), we obtain

$$f_s = ma\cos\theta - mg\sin\theta. \tag{4.25}$$

Substituting this equation and Eq. (4.23) into the condition that the object is yet to slide upward:

$$f_s \leq \mu_s N, \tag{4.26}$$

we attain the following inequality

$$ma\cos\theta - mg\sin\theta \leq \mu_s (ma\sin\theta + mg\cos\theta) \tag{4.27}$$

which we can solve for a as follows:

$$a \leq \frac{g\sin\theta + \mu_s g\cos\theta}{\cos\theta - \mu_s \sin\theta}. \tag{4.28}$$

This inequality indicates that

$$a_{max} = \frac{g\sin\theta + \mu_s g\cos\theta}{\cos\theta - \mu_s \sin\theta}. \tag{4.29}$$

From Eq. (4.27), we can also state the condition on μ_s at which the block remains stationary and not slide upward:

$$\mu_s \geq \frac{a\cos\theta - g\sin\theta}{a\sin\theta + g\cos\theta}. \tag{4.30}$$

This condition allows us to yield a limiting μ_s^l beyond which no amount of a can cause the block to slide upward. We can determine this limiting μ_s^l by letting $a \to \infty$, which leads to

$$\frac{a\cos\theta - g\sin\theta}{a\sin\theta + g\cos\theta} \to \frac{a\cos\theta}{a\sin\theta} = \cot\theta.$$

Since

$$\frac{a\cos\theta - g\sin\theta}{a\sin\theta + g\cos\theta} < \frac{a\cos\theta}{a\sin\theta},$$

we have

$$\mu_s^l = \cot\theta. \tag{4.31}$$

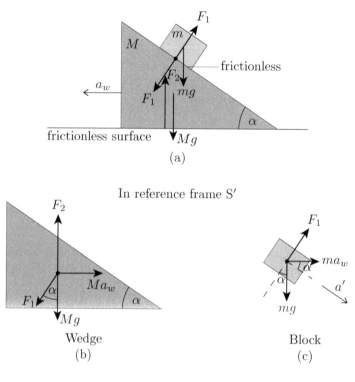

In reference frame S'

Wedge
(b)

Block
(c)

Fig. 4.13

4.10.2 *Object Sliding Down an Accelerating Wedge*

In this example, a wedge of mass M and angle of inclination α is initially at rest on a frictionless horizontal ground. A block of mass m is then placed on the frictionless surface of the wedge as shown in Fig. 4.13. The block slides down the incline from rest, and as it does so, the wedge is observed to accelerate to the left with a constant acceleration a_w (i.e., $\vec{a}_w = -a_w \hat{\imath}$). Our aim in this example is to determine a_w, and also the acceleration a of the block relative to the ground, from the perspective of the non-inertial reference frame. With this purpose in mind, we first attach our frame of reference S' to the wedge such that the wedge is stationary from the point of view of the observer. Since the wedge is accelerating with respect to the inertial reference frame of the ground, S' is a non-inertial reference frame, and because of which, application of Newton's second law needs to include fictitious forces. The direction of these fictitious forces is opposite to the direction of the frame's acceleration. For our example, there are two fictitious forces. One of which, Ma_w, acts on the wedge, while the other, ma_w, acts on the block. There is an interaction force F_1 between the block and the wedge. Because there is no friction between them, F_1 is perpendicular to the plane of their surfaces and act in opposite direction on each of them according to Newton's third law.

We begin our analysis on the wedge. Only the component equation of the forces in the horizontal direction on the wedge is relevant:

$$F_1 \sin \alpha = Ma_w \,, \tag{4.32}$$

since the equation of the vertical component of the forces does not serve our purpose. For the block, we have to consider the equation of motion of both components, although in this case, we treat them along the direction parallel and perpendicular to the contact surfaces of the two bodies:

$$F_1 + ma_w \sin \alpha = mg \cos \alpha \,, \tag{4.33}$$

$$ma' = mg \sin \alpha + ma_w \cos \alpha \,. \tag{4.34}$$

Notice that in both cases, fictitious forces have been added in our analysis as we apply Newton's second law of motion. From Eq. (4.33), we obtain

$$F_1 = mg \cos \alpha - ma_w \sin \alpha \,.$$

By substituting this equation into Eq. (4.32), we arrive at

$$a_w = \frac{mg \sin \alpha \cos \alpha}{M + m \sin^2 \alpha} \,, \tag{4.35}$$

$$= \frac{mg}{D} \,, \tag{4.36}$$

where

$$D = (M + m) \tan \alpha + \frac{M}{\tan \alpha} \,. \tag{4.37}$$

Note that Eq. (4.36) is obtained after we have performed trigonometric manipulations on Eq. (4.35). From Eq. (4.34), we obtain

$$a' = g \sin \alpha + a_w \cos \alpha \,,$$

$$= g \sin \alpha + \frac{mg \cos \alpha}{D} \,. \tag{4.38}$$

Let us now consider the x-component of the acceleration of the block relative to S':

$$a'_x = a' \cos \alpha \,,$$

$$= g \sin \alpha \cos \alpha + \frac{mg \cos^2 \alpha}{D} \,,$$

$$= \frac{Mg + mg}{D} \,. \tag{4.39}$$

In order to determine a, we first employ the relative acceleration relation:

$$\vec{a}_{B/O} = \vec{a}_{B/O'} + \vec{a}_{O'/O}.$$

It enables us to deduce the component form of the relative acceleration as follows:

$$a_x\hat{i} + a_y\hat{j} = a'_x\hat{i} + a'_y\hat{j} - a_W\hat{i},$$
$$= (a'_x - a_W)\,\hat{i} + a'_y\hat{j}. \tag{4.40}$$

By comparing between the components of the above equation, we have

$$a_x = a'_x - a_w,$$
$$= \frac{Mg + mg}{D} - \frac{mg}{D}.$$
$$= \frac{Mg}{D} \tag{4.41}$$

and

$$a_y = a'_y,$$
$$= -a'\sin\alpha,$$
$$= -\left(g\sin^2\alpha + \frac{mg\sin\alpha\cos\alpha}{D}\right),$$
$$= -\frac{(M+m)\,g\tan\alpha}{D}. \tag{4.42}$$

In other words,

$$\vec{a} = a_x\hat{i} + a_y\hat{j} = \frac{Mg}{D}\hat{i} - \frac{(M+m)\,g\tan\alpha}{D}\hat{j}. \tag{4.43}$$

Finally, let us make a comment on the results. When M is large, it is easy to deduce from Eq. (4.36) that $a_w \to 0$, which is to be expected. Furthermore, as $M \to \infty$, Eq. (4.43) implies that

$$\vec{a} = g\sin\alpha\,(\cos\alpha\hat{i} - \sin\alpha\hat{j}). \tag{4.44}$$

This shows that in the limit of large M, $|\vec{a}| = g\sin\alpha$ and the path of the block follows a straight line with gradient $-\tan\alpha$. These are again results consistent with our expectation.

Chapter 5

Energy and Work

5.1 Introduction

Like force, the physical concept of energy is of fundamental importance in mechanics. Energy gives the notion of a physical quantity that can be stored with the potentiality of generating motion. It can come in the form of potential energy, radiant energy, chemical energy, wave energy, matter energy, elastic energy, thermal energy, and when these are converted to the motive form, it becomes kinetic energy. One of the most important and fundamental principles in physics is the *conservation of energy*. This principle states that energy can neither be created nor destroyed, and its total amount remains constant even when converted from one form to another. It is a universal concept that is applicable to all processes in the universe, extending beyond the physical and chemical systems to the realm of biological organisms and engineering systems. Energy thus provides a powerful theoretical perspective on the physical processes that happen in the universe. It gives an alternative framework to the solutions of physical problems, and is especially useful in this respect when the forces concerned are not constant and vary across space.

5.2 System and Environment

While the seed of energy was planted by Gottfried Leibniz in the 17th century, the concept of energy is only fully developed in the 19th century within the science of thermodynamics. In fact, the first law of thermodynamics encapsulates the principle of conservation of energy.

Based on the perspective of thermodynamics, the universe can be divided into two parts: the system and the environment. These two parts

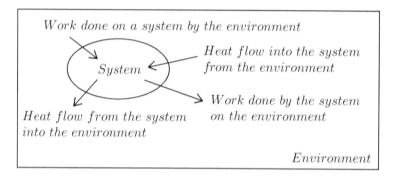

Fig. 5.1

are segregated by the system boundary (see Fig. 5.1). While the system can consist of a single particle, a collection of particles, or even a region of space, the environment contains physical entities that correspond to the rest of the universe whose details may be ignored during our analysis. Energy can flow from the environment to the system, or from the system into the environment. Such energy transfer is mediated by two physical processes: work done and heat flow. When energy is transferred from the environment to the system, this could be due to work done on the system by the environment and/or heat flow into the system from the environment. Conversely, when energy is transferred from the system to the environment, it could result from work done by the system on the environment and/or heat flow from the system into the environment. Essentially, it is important to note that energy is conserved during the transfer process, and it can convert from one form to another during the process of transfer.

5.3   Work

Let us develop the concept of energy through the quantitative definition of work. It is defined as follows:

The work W done on a system by an external agent exerting a constant force on the system is the scalar product of the force $\vec{F}$ and the displacement $\Delta\vec{r}$ of the point of application of the force:

$$W = \vec{F} \cdot \Delta\vec{r} = |\vec{F}||\Delta\vec{r}|\cos\theta\,, \qquad (5.1)$$

where θ is the angle between the force and the displacement vector (see Fig. 5.2).

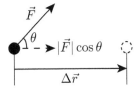

$$\vec{F}$$
$$\theta$$
$$|\vec{F}|\cos\theta$$
$$\Delta\vec{r}$$

Fig. 5.2

Work thus quantifies the effectiveness of a force in mediating the transfer of energy to or from a system and its environment. Note that work is a scalar quantity and its unit is the joule (J) or Newton meter (N.m).

Let us consider an object resting on a frictionless surface being acted on by a horizontal force $\vec{F}_h$, a normal force $\vec{N}$, and its weight $\vec{W}$ as shown in Fig. 5.3(a). The object is observed to slide horizontally with a displacement $\Delta\vec{r}$. It is easy to deduce that the work done by $\vec{N}$ and $\vec{W}$ is zero since the angle between them and the displacement is 90° (because $\cos 90° = 0$). On the other hand, the work done by $\vec{F}$ is $|\vec{F}_h||\Delta\vec{r}|$ and is non-zero. The idea is that when the acting force and the displacement of the point of application of the force is perpendicular, the force is not effective in transferring any energy at all. This is further illustrated in Fig. 5.3(b) with regards to the action of the centripetal force on a body under uniform circular motion. The fact that the centripetal force is perpendicular to the displacement of the body at all time indicates that it serves only to change the direction of the body without enhancing its motion. Hence, there is no energy transfer. On the other hand, the force $\vec{F}_h$ of Fig. 5.3(a) does act to increase the speed of the object and thus enhances its motion, and there is a positive transfer of energy from the external agent (environment) exerting $\vec{F}_h$ to the object (system). Similarly, Fig. 5.3(c) shows a truck accelerating from rest and in consequence causes the static friction $\vec{F}_s$ to act on its load. As the truck and load displaces by $\Delta\vec{r}$, an energy of $|\vec{F}_s||\Delta\vec{r}|$ is transferred to the load (system) through the positive work done by $\vec{F}_s$, which is exhibited by an increase in speed of the load. This energy comes from the truck (environment) and can be traced to the chemical energy of its diesel fuel. On the other hand, Fig. 5.3(d) shows a mass being impeded by the kinetic friction $\vec{F}_k$ during its motion. As the mass displaces by $\Delta\vec{r}$, the kinetic friction performs a work of $-|\vec{F}_k||\Delta\vec{r}|$ on the mass. Notice the work done is negative, which means that the kinetic friction has taken the amount of energy $|\vec{F}_k||\Delta\vec{r}|$ from the mass (system) and passed it to the environment in the form of heat.

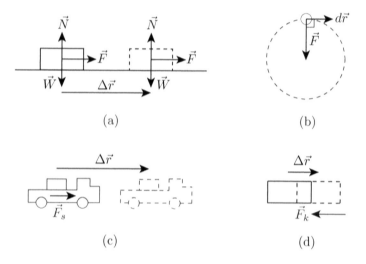

Fig. 5.3

In summary, when work is done by the environment on the system, i.e. energy is transferred from the environment to the system, *work is positive*. On the other hand, when work is done by the system on the environment, i.e. energy is transferred from the system to the environment, *work is negative*. The manner in which work is done depends on the relative direction between the applied force and the displacement. Moreover, the effectiveness in energy transfer in each case of positive and negative work done depends on the degree of alignment between the acting force and the displacement of its point of application. A system, by interacting with its environment, transfers energy across the system boundary. Based on the principle of conservation of energy, if the amount of energy stored in the system changes by ΔE, the amount of energy stored in the environment will change by $-\Delta E$, and vice versa.

In general, the force that acts on a body need not be a constant. However, by considering the force to be a constant within an infinitesimal displacement of $d\vec{r}$, and adding up all the consecutive work done $\vec{F} \cdot d\vec{r}$ to accumulate the resulting sum, we obtain the following more general definition of work for a single force:

$$W = \int \vec{F} \cdot d\vec{r}. \tag{5.2}$$

If more than one force acts on the system and the system can be modeled as a particle, the total work done on the system is the work done by the net force:

$$\sum W = W_{\text{net}} = \int \sum_i \vec{F}_i \cdot d\vec{r}. \tag{5.3}$$

However, if the system cannot be modeled as a particle, then the total work is the algebraic sum of the work done by the individual forces.

5.4 Hooke's law and the Elastic Spring

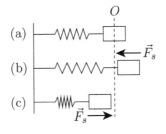

Fig. 5.4

Figure 5.4 shows the configuration of a spring and a mass m attached to it. When the mass is at the origin O, the spring is neither stretched nor compressed. There is no net force acting on the mass and O is known as the equilibrium position. As the mass is being pulled in the positive x direction, a net elastic force from the spring acts on the mass to the left towards O as shown in Fig. 5.4(b). If the mass is being pushed in the negative x direction, the net elastic force now exerts rightwards towards O (see Fig. 5.4(c)). In both cases, we notice that the elastic force $\vec{F}_s$ always acts in a direction that is opposite to the displacement $\vec{x}$ of the mass from O. Specifically, $\vec{F}_s$ and $\vec{x}$ follow the linear relationship:

$$\vec{F}_s = -k\vec{x}, \tag{5.4}$$

where k is the spring constant which measures the stiffness of the spring. Equation (5.4) is known as Hooke's law.

5.5   Energy Transfer in Spring-Mass System

5.5.1   *Work Done by Applied Force on Spring-Mass System*

Let us begin by first assuming that the spring plus mass is the system and the external agent is the environment. The agent applied a force $\vec{F}_{\mathrm{app}}$ on the mass in a quasi-static fashion to stretch the spring to a maximum extension of x_{max}. *Quasi-static* here means that the mass is "almost" not moving during the energy transfer by the external agent. In order to do that, $\vec{F}_{\mathrm{app}}$ must counteract the spring force given by Eq. (5.4) at each point in time. Hence, the magnitude of $|\vec{F}_{\mathrm{app}}| = |k\vec{x}| = kx$. Then, the work done on the spring-mass system by the external agent is given by:

$$
\begin{aligned}
W &= \int \vec{F}_{\mathrm{app}} \cdot d\vec{x} \\
&= \int |\vec{F}_{\mathrm{app}}|\, |d\vec{x}| \\
&= \int_0^{x_{\mathrm{max}}} kx\, dx \\
&= \frac{1}{2}kx_{\mathrm{max}}^2 .
\end{aligned}
\tag{5.5}
$$

Note that $\vec{F}_{\mathrm{app}}$ is in the same direction of the displacement $d\vec{x}$ of the mass (see Fig. 5.5). Thus, the amount of energy transferred from the external agent to the spring-mass system is $kx_{\mathrm{max}}^2/2$. In other words, the spring-mass system has gained an energy of $kx_{\mathrm{max}}^2/2$.

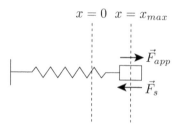

Fig. 5.5

While the external agent exerts a force of $\vec{F}_{\mathrm{app}}$ on the spring-mass system, the spring-mass system exerts an equal and opposite force of $\vec{F}'_{\mathrm{app}} = -\vec{F}_{\mathrm{app}}$ on the external agent corresponding to Newton's third law.

Consequently, the work done on the external agent by the spring-mass system is

$$W' = \int \vec{F}'_{app} \cdot d\vec{x}$$

$$= -\int \vec{F}_{app} \cdot d\vec{x}$$

$$= -\frac{1}{2} k x_{max}^2. \tag{5.6}$$

This implies that the amount of energy transferred from the spring-mass system to the external agent is $-k x_{max}^2/2$. In other words, the external agent has lost an energy of $k x_{max}^2/2$ while the spring-mass system has gained this same amount of energy. Notice that this equal gain and loss results from the action and reaction forces of Newton's third law, which mediate the transfer as the two bodies interact with each other. The fact that the action and reaction forces are equal in magnitude but opposite in direction does serve an important role in the manner the energy is transmitted between the two bodies.

5.5.2 *Quasi-static Transfer of Energy*

While $k x_{max}^2/2$ of energy has been transferred to the spring-mass system in a quasi-static manner, where does the energy go? To answer this question, let us analyze again by treating the mass as the system, but the hand and the spring now as the environment.

The work done on the mass (system) by the external agent (environment) is:

$$\int \vec{F}_{app} \cdot d\vec{x} = \int_0^{x_{max}} k x \, dx$$

$$= \frac{1}{2} k x_{max}^2. \tag{5.7}$$

This is also the energy transfer from the agent to the mass. Next, the work done on the mass (system) by the spring (environment) is:

$$\int \vec{F}_s \cdot d\vec{x} = -\int \vec{F}_{app} \cdot d\vec{x}$$

$$= -\frac{1}{2} k x_{max}^2, \tag{5.8}$$

which is the energy transfer from the spring to the mass. In fact, the negative sign informs us that the mass has transferred an amount of $kx^2_{\text{max}}/2$ to the spring.

On the whole, the mass has gained zero amount of energy. The $kx^2_{\text{max}}/2$ amount of energy from the external agent has been transferred to the spring. This energy is stored as elastic potential energy in the spring (see definition of potential energy in the next section).

5.5.3   Spring-Mass System as an Isolated System

With the spring-mass system now possessing $kx^2_{\text{max}}/2$ amount of energy, let us now consider the interaction between mass and spring without any external agent. In this context, the spring-mass system becomes an isolated system. Nonetheless, we may still treat the mass as the system and the spring as the environment or vice versa, in our analysis. Then, the work done on the mass by the spring is as follows:

$$
\begin{aligned}
W_{\text{spring-on-mass}} &= \int \vec{F}_s \cdot d\vec{x} \\
&= \int m\vec{a} \cdot d\vec{x} \\
&= \int m\frac{d\vec{v}}{dt} \cdot d\vec{x} \\
&= m \int \frac{d\vec{v}}{dt} \cdot \vec{v}\, dt \\
&= m \int \frac{1}{2}\frac{d}{dt}\left(v^2\right) dt \\
&= \frac{1}{2}m \int_{v_i^2}^{v_f^2} d\left(v^2\right) \\
&= \frac{1}{2}mv_f^2 - \frac{1}{2}mv_i^2 .
\end{aligned} \tag{5.9}
$$

Note that the fifth line of Eq. (5.9) results from a series of mathematical manipulations. First, $d(\vec{v} \cdot \vec{v})/dt = d\vec{v}/dt \cdot \vec{v} + \vec{v} \cdot d\vec{v}/dt$. Since dot product commutes, this implies $d(v^2)/dt = 2d\vec{v}/dt \cdot \vec{v}$, which leads to $d\vec{v}/dt \cdot \vec{v} = (1/2)d(v^2)/dt$.

Evidently, the above work done can also be evaluated in the following manner:

$$
\begin{aligned}
W_{\text{spring-on-mass}} &= \int \vec{F}_s \cdot d\vec{x} \\
&= \int -k\vec{x} \cdot d\vec{x}
\end{aligned}
$$

$$= \int -kx\hat{\imath} \cdot dx\hat{\imath}$$

$$= -k \int_{x_i}^{x_f} x \, dx$$

$$= \left[-\frac{1}{2}kx^2 \right]_{x_i}^{x_f}$$

$$= \frac{1}{2}kx_i^2 - \frac{1}{2}kx_f^2 \tag{5.10}$$

Equating Eq. (5.9) and Eq. (5.10), we obtain

$$\frac{1}{2}mv_i^2 + \frac{1}{2}kx_i^2 = \frac{1}{2}mv_f^2 + \frac{1}{2}kx_f^2 . \tag{5.11}$$

Equation (5.11) is a statement of the Work-Energy Theorem. This theorem serves to define two quantities. The first quantity is the kinetic energy

$$K = \frac{1}{2}mv^2 , \tag{5.12}$$

which measures the motive energy of the mass through its velocity. The second quantity defines the potential energy

$$U = \frac{1}{2}kx^2 . \tag{5.13}$$

Potential energy is the energy stored in the configuration of the system. The system in this case is the spring, with the configuration being given by the displacement of the spring x from its equilibrium position. From these definitions, Eq. (5.11) expresses the principle of mechanical energy conservation of an isolated system, i.e. the total kinetic and potential energy of the isolated spring-mass system is conserved and remains constant.

Returning to the initial state when the spring-mass system just receives an energy of $kx_{\text{max}}^2/2$ from the external agent, we note that at that instant, the mass is not moving. All the energy is stored as elastic potential energy in the spring, which is maximally stretched with displacement x_{max} from the equilibrium position. As the spring force does work on the mass, it transfers elastic potential energy of the spring to kinetic energy of the mass. The transfer is done gradually, with more and more potential energy being converted to kinetic energy as the mass approaches the equilibrium point of the spring. At the equilibrium point, all the $kx_{\text{max}}^2/2$ of potential energy has been transformed to kinetic energy (i.e., $kx_{\text{max}}^2/2 = mv_{\text{max}}^2/2$). The kinetic energy of the mass reaches a maximum at this point: $mv_{\text{max}}^2/2$.

5.5.4   *Dissipative Interaction of Spring-Mass System with the Environment*

Previously, we have considered the situation where the energy flows in from the external environment to the spring-mass system, and the case of an isolated spring-mass system with mechanical energy interchanging internally between the mass and the spring. In this section, we shall look at the interaction between the spring-mass system and the external environment through the action of a dissipative force.

The dissipative force concerned is the frictional force. Let us place the spring-mass system horizontally on the ground such that the mass slides on its surface (to be treated as the environment) with friction. Due to sliding, the friction encountered by the mass is the kinetic friction f_k. The work done by the surface on the mass is $-f_k d$, where d is the distance traveled by the mass. The presence of the negative sign indicates that the mass travels in opposite direction to the action of the kinetic friction. Moreover, the negative sign indicates that energy is lost from the mass to the surface. Because the spring interacts with the mass at the same time, the mass continually draws its kinetic energy from the potential energy of the spring. Thus, the loss of mechanical energy to the surface also affects the quantity of potential energy stored in the spring. As a whole, we expect the total potential energy and kinetic energy of the spring-mass system to reduce by $-f_k d$. We express this change in energy ΔE as a corresponding change in kinetic energy ΔK and potential energy ΔU as follows:

$$\Delta E = \Delta K + \Delta U = -f_k d \,. \tag{5.14}$$

5.6   Potential Energy and Conservative Force

There is an intricate relation between the potential energy function $U(x)$ and the force $F_C(x)$ that it is associated with. This relationship can be understood from the following argument.

Assume that a mass interacts with F_C. During the interaction, the force does work on the mass, leading to a transfer of an amount of energy $F_C \Delta x$ to the mass as the mass moves from x to $x + \Delta x$. Since energy is conserved, this energy must come from a stored energy, called the potential energy associated with the force. This stored, or potential, energy is quantified by the potential energy function. Specifically, when the mass is at x, the associated potential energy function is $U(x)$. As it moves to $x + \Delta x$, the associated potential energy function $U(x + \Delta x) < U(x)$ because it has supplied $F_C \Delta x$ amount of energy to the mass as explained above. In consequence of this energy transfer, we have

$$U(x + \Delta x) - U(x) = -F_C \Delta x .$$

If we take the infinitesimal limit of small displacement, i.e., $\Delta x \to 0$, we obtain the following relationship between F_C and $U(x)$:

$$\begin{aligned} F_C &= - \lim_{\Delta x \to 0} \frac{U(x + \Delta x) - U(x)}{\Delta x} \\ &= -\frac{dU}{dx} . \end{aligned} \tag{5.15}$$

Equation (5.15) defines F_C to be a conservative force. It tells us that associated with a conservative force is a corresponding potential energy function. For example, the potential energy $U_s(x)$ that is linked to the spring force $F_s = -kx$ can be determined as follows:

$$\begin{aligned} -\frac{dU_s}{dx} &= F_s \\ &= -kx \end{aligned}$$

or

$$\begin{aligned} U_s(x) &= \int kx \, dx \\ &= \frac{1}{2}kx^2 + C , \end{aligned}$$

where C is an arbitrary constant. For the gravitational force $F_g = -mg$, a similar evaluation leads to $U_g(x) = mgx + C$. The presence of the constant C implies that the choice of the reference for the potential energy function $U(x)$ is arbitrary. In other words, only the difference in potential energy is physically meaningful.

Conservative force obeys two equivalent properties: (a) the work done by a conservative force on a particle moving between any two points is independent of the path taken by the particle; (b) the work done by a conservative force on a particle moving through any closed path is zero. Let us illustrate how the spring force fulfills these two properties. We first label O to be the position $x = 0$ where the particle is at the equilibrium point of the spring. We label A (B) to be the position of maximum extension (compression) of the spring with the particle at $x = x_{max}$ ($x = -x_{max}$). We define path I to be from O $\to$ A; path II from O $\to$ A $\to$ O $\to$ B $\to$ O $\to$ A; and path III from O $\to$ A $\to$ O. Although path I and II are different paths, the particle moves between the same two points O and A in both paths. On the other hand, path III is a closed path as the beginning and end points are the same point O. The work done by the spring force on the particle is the same $U_s(x_{\max}) - U_s(0) = U_s(x_{\max})$ for path I and II, and hence is independent of path as per property (a). For path III, the work done by the spring force on the particle is $U_s(0) - U_s(0) = 0$, which fulfills property (b) of a conservative force. In fact, it is easy to check that the spring force fulfills both property (a) and (b) by taking other points in the spring-mass motion.

On the other hand, a nonconservative force does not satisfy property (a) and (b) of conservative forces. A nonconservative force is the kinetic frictional force F_k. For path I, the work done by the kinetic friction is $-F_k x_{\max}$. But for path II, it is $-5F_k x_{\max}$. Thus, the work done depends on path even though the start and end points are the same in both cases, which implies that property (a) is not fulfilled. For path III, the work done is $-2F_k x_{\max}$, which is not zero and hence, property (b) is also not obeyed. The reason that frictional force is nonconservative arises from the fact that it is a many-body effective force as discussed in section 4.6. While the individual forces acting between the atoms on the surfaces of the particle and the ground are conservative, their *collective effect as a force* as represented by friction is nonconservative. It is related to thermodynamic irreversibility with the generation of heat, and is caused by random interactions of a large ensemble of atoms between the two bodies.

5.7   Collision of Two Identical Mass via a Spring-like Force

Let us consider a mass m moving with velocity u towards another identical mass which is at rest with an elastic spring of spring constant k attached as shown in Fig. 5.6 from the perspective of an inertial reference frame S. We will determine the motion of the two masses as the first mass presses on the spring until both of them move at the same velocity. We will also deduce the energy transfer processes and outcome during the interaction between the masses via the spring. We assume there is no loss of energy to friction or resistive forces throughout the process.

It is convenient to take the perspective from another inertial reference frame S' that moves at a velocity $u/2$ to the right of S to describe the dynamical interaction between the two masses. Based on S', the two masses interact in the manner depicted in Fig. 5.7. By experience, we know that

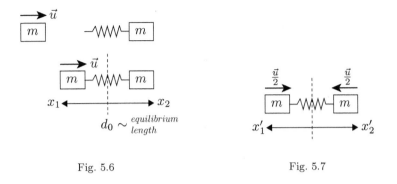

Fig. 5.6   Fig. 5.7

the motion of the two masses is simple harmonic, such that

$$\frac{dx'_1}{dt} = \frac{u}{2} \cos \omega t, \tag{5.16}$$

$$\frac{dx'_2}{dt} = -\frac{u}{2} \cos \omega t, \tag{5.17}$$

where x'_1 and x'_2 are the displacement of mass 1 and 2 from the origin in frame S'.

Transforming back to the original frame S, their velocities are given by the following equations:

$$v_1 = \frac{u}{2} + \frac{u}{2} \cos \omega t, \tag{5.18}$$

$$v_2 = \frac{u}{2} - \frac{u}{2} \cos \omega t. \tag{5.19}$$

Differentiating Eqs. (5.18) and (5.19) with respect to time, we arrive at their accelerations (which should be the same in both inertial reference frames) as follows:

$$a_1 = -\tfrac{\omega u}{2} \sin \omega t, \tag{5.20}$$

$$a_2 = \tfrac{\omega u}{2} \sin \omega t. \tag{5.21}$$

Integrating Eqs. (5.18) and (5.19) with respect to time, we obtain for the displacement of the first mass:

$$x_1 = \frac{u}{2} t + \frac{u}{2\omega} \sin \omega t + C_1,$$

where C_1 is an arbitrary constant. When $t = 0$, $x_1 = -d_0/2$. Therefore, $C_1 = -d_0/2$. Similarly,

$$x_2 = \frac{u}{2}t + \frac{u}{2\omega}\sin\omega t + C_2 \,.$$

When $t = 0$, $x_2 = d_0/2$. In this case, $C_2 = d_0/2$. In summary, we have the following set of displacement equations for the two masses:

$$x_1 = \frac{u}{2}t + \frac{u}{2\omega}\sin\omega t - \frac{d_0}{2} \,, \tag{5.22}$$

$$x_2 = \frac{u}{2}t - \frac{u}{2\omega}\sin\omega t + \frac{d_0}{2} \,. \tag{5.23}$$

Note that we can also analyze the dynamics as displayed in Fig. 5.6 by considering the extension of the spring:

$$\Delta x(t) = d_0 - [x_2(t) - x_1(t)] \,,$$

with $\Delta x(0) = 0$. If $x_2(t) - x_1(t) < d_0$, which implies that $\Delta x > 0$, the spring is compressed, and mass 2 will accelerate to the right while mass 1 to the left. These physical conditions can be expressed as follows:

$$ma_2 = k\Delta x = k\left\{d_0 - [x_2(t) - x_1(t)]\right\} \,,$$
$$ma_1 = -k\Delta x = -k\left\{d_0 - [x_2(t) - x_1(t)]\right\} \,,$$

or

$$a_1 = -\frac{k}{m}\left\{d_0 - [x_2(t) - x_1(t)]\right\} \,, \tag{5.24}$$

$$a_2 = \frac{k}{m}\left\{d_0 - [x_2(t) - x_1(t)]\right\} \,. \tag{5.25}$$

Substituting Eqs. (5.22) and (5.23) into Eq. (5.25), we have

$$a_2 = \frac{k}{m}\left\{d_0 - \left[-\frac{u}{\omega}\sin\omega t + d_0\right]\right\} = \frac{ku}{m\omega}\sin\omega t \,.$$

Comparing this equation against Eq. (5.21), we have

$$\frac{ku}{m\omega} = \frac{\omega u}{2} \implies \omega = \sqrt{\frac{2k}{m}} \,.$$

Let us now examine the energy transfer between mass 1, mass 2 and the spring from $t = 0$ to $t = \pi/2\omega$. Note that at $t = \pi/2\omega$, the two masses travel at the same velocity: $v_1 = v_2 = u/2$. At $t = 0$, $v_1 = u$, $v_2 = 0$, and $\Delta x = 0$. Thus, the total initial energy of the system is just equal to the kinetic energy of mass 1:

$$K_i = \frac{1}{2}mv_1^2 = \frac{1}{2}mu^2. \tag{5.26}$$

At $t = \pi/2\omega$, $v_1 = u/2$ and $v_2 = u/2$. Thus, the total kinetic energy of the system is

$$\begin{aligned} K_f &= \frac{1}{2}mv_1^2 + \frac{1}{2}mv_2^2, \\ &= \frac{1}{2}m\left(\frac{u}{2}\right)^2 + \frac{1}{2}m\left(\frac{u}{2}\right)^2, \\ &= \frac{1}{4}mu^2. \end{aligned} \tag{5.27}$$

Next, we will determine the extension Δx at $t = \pi/2\omega$. From Eqs. (5.22) and (5.23), we first obtain the displacement of mass 1 and 2 at $t = \pi/2\omega$:

$$x_1 = \frac{u\pi}{4\omega} + \frac{u}{2\omega} - \frac{d_0}{2},$$

$$x_2 = \frac{u\pi}{4\omega} - \frac{u}{2\omega} + \frac{d_0}{2}.$$

Therefore,

$$\begin{aligned} \Delta x\left(\frac{\pi}{2\omega}\right) &= d_0 - \left[x_2\left(\frac{\pi}{2\omega}\right) - x_1\left(\frac{\pi}{2\omega}\right)\right], \\ &= \frac{u}{\omega}. \end{aligned} \tag{5.28}$$

The elastic potential energy of the spring at $t = \pi/2\omega$ is

$$\begin{aligned} U_e &= \frac{1}{2}k\Delta x^2, \\ &= \frac{1}{2}k\left(\frac{u}{\omega}\right)^2, \\ &= \frac{1}{2}ku^2\left(\frac{m}{2k}\right), \\ &= \frac{1}{4}mu^2. \end{aligned} \tag{5.29}$$

These results show that at $t = \pi/2\omega$, the conservation of energy applies as follows:

$$K_i = U_e + K_f. \tag{5.30}$$

Interestingly, unlike the spring-mass case considered in section 5.5 where the mass and spring are all acted by forces of an action and reaction pair with equal and opposite magnitude, the transfer of energy here has not led to an equivalence between initial kinetic energy and final kinetic energy alone. This results from the difference in displacement in the work done against the masses as shown below:

$$\Delta x_1 = x_1\left(\frac{\pi}{2\omega}\right) - x_1(0) = \frac{u\pi}{4\omega} + \frac{u}{2\omega}, \tag{5.31}$$

$$\Delta x_2 = x_2\left(\frac{\pi}{2\omega}\right) - x_2(0) = \frac{u\pi}{4\omega} - \frac{u}{2\omega}. \tag{5.32}$$

This difference in displacement results from the different speeds of the two masses, leading to the rest of the energy from mass 1 being deposited as elastic potential energy in the elastic spring. In fact, this physical situation is analogous to the perfectly inelastic collision encountered in collision problems.

5.8 Pseudowork

So far, we have discussed on the transfer of energy between system and environment that crosses the system boundary through the process of work done. Recall that such a process happens when the scalar product of the force and the displacement of its point of application is nonzero. For example, in the work done by kinetic friction, the point of application of the frictional force displaces along the surface of the object that it acts, and hence the resultant work done on the object is nonzero. In this case, energy flows out of the object (which is the system), to the floor surface (which is the environment) on which the object slides.

Now, it is possible to define another type of work done which does not result in any flow of energy across the system boundary, but instead it mediates a transfer of energy from one mode to another internal to the system. Let us take a running person who sprints from rest as an example (see Fig. 5.8) to illustrate this new type of work done. Reasoning on physical grounds, a running person accelerates as a consequence of the action of static friction between its shoes and the floor surface. By the application of Newton's second law, we oberve that $2f = Ma$, where f is the static friction between one of the person's shoe and the floor surface, M is the mass of the person, and a is its acceleration. Notice that the static friction f does no work because its point of application has zero displacement as the person runs. Nonetheless, it serves the purpose of mediating a transfer of the chemical energy of the person to its kinetic energy of translation. Without static friction, the person would slip and there is no forward movement.

Fig. 5.8

To account for such internal transfer of energy within a system motivates the definition of *pseudowork*[1]. Pseudowork is a form of work done which mediates a transfer of energy from one mode to another within a system, with at least one of the mode being kinematic in nature. It is defined as the scalar product of the force $\vec{F}$ and the displacement of the center of mass[2] $\Delta \vec{r}_{CM}$ of the object, while the *real* work done by the force $\vec{F}$ is zero. Specifically, pseudowork (W_{pseudo}) is defined as follows:

$$W_{pseudo} = \vec{F} \cdot \Delta \vec{r}_{CM} = |\vec{F}||\Delta \vec{r}_{CM}| \cos \phi \qquad (5.33)$$

and

$$W_{real} = \vec{F} \cdot \Delta \vec{r} = 0 \,, \qquad (5.34)$$

where ϕ is the smaller angle between $\vec{F}$ and $\Delta \vec{r}_{CM}$, W_{real} is the real work done by $\vec{F}$, and $\Delta \vec{r}$ is the displacement of the point of application of $\vec{F}$. Based on this definition, the static friction in the above example performs a pseudowork of

$$f d_{CM} = \frac{1}{2} v_{CM,f}^2 - \frac{1}{2} v_{CM,i}^2 \qquad (5.35)$$

that allows the person to translate as it runs. Through the pseudowork of static friction, the center of mass of the person displaces by a distance of d_{CM} as its velocity increases from $v_{CM,i}$ to $v_{CM,f}$. No energy flows out of the system of the person to the environment because static friction performs zero real work. On the other hand, in the kinetic friction example discussed earlier, there is no pseudowork since there is a flow of energy across the system boundary. More examples on the application of pseudowork will be given in Chapter 7.

---

[1]B. A. Sherwood, "Pseudowork and real work", *Am. J. Phys.* **51**(7), 597 (1983).
[2]The definition and details of the center of mass of a system will be covered in Chapter 6.

Chapter 6

Linear Momentum and Newton's Law with Variable Masses

6.1 Introduction

In conceiving a quantifier of motion, Newton had defined a basic quantity termed *linear momentum*. He used this quantity to formulate his second law of motion, where he defined force as the rate of change of linear momentum. Furthermore, linear momentum is also the key notion in the conceptualization of a general principle in physics: the *Conservation of Linear Momentum*. This principle finds utility in collision problems and the physics of many-body interacting systems. It forms a fundamental law in nature that complements the law of conservation of energy, with both having wide applicability in all branches of physics. In this chapter, we shall cover these foundational aspects of linear momentum, as well as to employ it to construct a general scheme to treat Newton's Second Law with variable masses.

6.2 Momentum and Newton's Laws of Motion

Newton defined linear momentum[1] as a fundamental quantity of motion that is equal to the product of the mass m and velocity $\vec{v}$ of a particle (or an object that can be modeled as a particle):

$$\vec{p} = m\vec{v}. \tag{6.1}$$

As such, momentum is a vector quantity with unit kg.m/s. Newton's second law of motion then states that

---

[1]We shall define a rotational version of momentum in the next chapter

*The time rate of change of the linear momentum of a particle is equal to
the net force acting on the particle:*

$$\sum \vec{F} = \frac{d\vec{p}}{dt} \,. \tag{6.2}$$

Notice that when the mass m is a constant, Eq. (6.2) becomes $\sum \vec{F} = d\,(m\vec{v})\,/dt = md\vec{v}/dt = m\vec{a}$, which is the version of Newton's second law as defined in Eq. (4.1) of Chapter 4.

Next, let us consider a *system* of two particles with mass m_1 and m_2 interacting with each other. The forces of interaction are the action and reaction pair: $\vec{F}_{1\to 2}$ and $\vec{F}_{2\to 1}$, between the two particles. These forces constitute the *internal forces* of the system. Since there are no forces that act from outside of this system (which are known as the *external forces*), the system is considered to be isolated.

According to Newton's third law, we have

$$\vec{F}_{2\to 1} + \vec{F}_{1\to 2} = 0 \,.$$

Applying Newton's second law as given by Eq. (6.2) to each particle with respective linear momentum $\vec{p}_1$ and $\vec{p}_2$, this equation becomes

$$\frac{d\vec{p}_1}{dt} + \frac{d\vec{p}_2}{dt} = 0 \,,$$

$$\frac{d\,(\vec{p}_1 + \vec{p}_2)}{dt} = 0 \,,$$

where the second line of the equation results from the linearity of the differentiation operation. This result implies that

$$\vec{p}_1 + \vec{p}_2 = \vec{c}, \tag{6.3}$$

where $\vec{c}$ is a constant vector. Equation (6.3) is a statement of *conservation of linear momentum* of the two interacting particles. It indicates that the total linear momentum (i.e. vector sum) of the isolated system of two particles is independent of time, although the individual linear momentum of each particle can depend on time. It is important to note in this case that total linear momentum continues to be conserved at the level of individual Cartesian components as follows:

$$p_{x_1} + p_{x_2} = c_x \,,$$

$$p_{y_1} + p_{y_2} = c_y \,,$$

$$p_{z_1} + p_{z_2} = c_z \,,$$

where c_x, c_y and c_z are constant in value and are the Cartesian components of $\vec{c}$.

6.3 Impulsive Force and the Collision Problem

Interaction between matter is a commonplace occurrence in physics and one unique form of such interaction is that of collision. While collision events can involve many particles, a typical event usually concerns two particles that come close to each other and interact by means of forces. The time interval of such two-particle collision is very short and as a result, the collision is heavily dominated by the force that the two particles interact with each other. It is in this context that the concept of an impulsive force possesses a distinctive flavor. Impulsive forces that occur in such collision events are internal to the system and may vary with time in a complicated way.

Before we clarify the concept of an impulsive force, let us first define *impulse*. The concept of impulse relates to a net force $\vec{F}$ that may vary with time. It originates from Newton's second law being reexpressed in the following way:

$$d\vec{p} = \vec{F}dt. \tag{6.4}$$

If the force $\vec{F}$ were to act over the time interval t_i to t_f on a particle such that the linear momentum of the particle changes from $\vec{p}_i$ to $\vec{p}_f$, we can integrate Eq. (6.4) to obtain:

$$\Delta\vec{p} = \vec{p}_f - \vec{p}_i = \int_{t_i}^{t_f} \vec{F}dt. \tag{6.5}$$

The quantity $\int_{t_i}^{t_f} \vec{F}dt$ is called the *impulse* $\vec{I}$ of the force $\vec{F}$ acting on the particle in a time interval $\Delta t = t_f - t_i$, i.e.,

$$\vec{I} = \int_{t_i}^{t_f} \vec{F}dt. \tag{6.6}$$

This gives rise to the Impulse-Momentum theorem, which states that

The impulse of the force $\vec{F}$ acting on a particle equals the change in the linear momentum of the particle:

$$\vec{I} = \vec{p}_f - \vec{p}_i. \tag{6.7}$$

Note that impulse is a vector, has the dimension of linear momentum, and it is not a force. The magnitude of the impulse $|\vec{I}|$ is equal to the area under the force-time curve. Furthermore, the impulse can be found through the time-averaged force $\vec{F}_{av}$, which gives the same impulse as the time-varying force $\vec{F}$ over the same time interval Δt:

$$\vec{I} = \vec{F}_{av}\Delta t \, . \tag{6.8}$$

In a collision, the duration of a force acting on the particle is very short. Such a force is known as the *impulsive force*. In many physical situations, there is only one impulsive force among the presence of other forces. Typically, this impulsive force is far greater in magnitude and act for a much shorter time than the other forces. Hence, in the analysis of collision problems, it is justified to ignore the other forces. For example, the impulsive force of contact between two vehicles in a collision is far greater than the frictional force between the vehicles and the road, such that the frictional force is ignored in the analysis. This assumption is known as the *impulsive approximation.*

During a two-particle collision, the particles interact with each other at close range, whereupon they move very little during the short time interval in which they interact. Because the interaction between them is dominated by the impulsive force which is internal to the system, we can ignore other external forces by employing the impulsive approximation. From this context, we can treat the two-particle collision system to be isolated and apply the principle of conservation of linear momentum. In other words, the total momentum is considered to be conserved (i.e. constant in time) in all such collisions.

Specifically, let us consider a collision between two particles of mass m_1 and m_2. Before the collision, the particles are traveling at velocities $\vec{v}_{1i}$ and $\vec{v}_{2i}$ respectively. After the collision, their respective velocities are $\vec{v}_{1f}$ and $\vec{v}_{2f}$. Note that the subscripts i and f denote "initial" and "final" respectively. With the impulsive approximation, the two-particle system at the *moments* of collision can be treated to be acted upon only by the internal collision force. Thus, the total linear momentum is conserved for the system *just* before and after the collision as follows:

$$\vec{p}_{1i} + \vec{p}_{2i} = \vec{p}_{1f} + \vec{p}_{2f} \tag{6.9}$$

or

$$m_1\vec{v}_{1i} + m_2\vec{v}_{2i} = m_1\vec{v}_{1f} + m_2\vec{v}_{2f} \, . \tag{6.10}$$

We shall now get into further details on the collision between the two particles. If the collision between the two particles conserves the total kinetic energy of the system, the collision is known as an *elastic collision.* During an elastic collision, there is no way for energy to be transformed to another form. Hence, the total kinetic energy of the two particles before and after the collision remain unchanged. Explicitly, we have

$$\frac{1}{2}m_1|\vec{v}_{1i}|^2 + \frac{1}{2}m_2|\vec{v}_{2i}|^2 = \frac{1}{2}m_1|\vec{v}_{1f}|^2 + \frac{1}{2}m_2|\vec{v}_{2f}|^2. \tag{6.11}$$

Such a physical condition happens during the collision between elementary particles such as protons, electrons, or neutrons. For macroscopic collisions between billiard balls, the collision can be considered approximately elastic.

However, there are cases when the energy can be transformed into other forms after the collision. Under these circumstances, the total kinetic energy of the two-particle system is not the same before and after the collision. Total kinetic energy is not conserved and the collision is known as an *inelastic collision*. Usually, the initial total kinetic energy is lost as heat and sound during inelastic collision. In the special case when the two particles stick together after the collision, it is a *perfectly inelastic collision*. Energy is further converted into stored internal energy in this case.

Regardless of whether the collision is elastic or inelastic, the total linear momentum of the system is always conserved. This condition gives rise to a number of equations that is equal to the number of dimensions considered in the collision problem. For example, in one-dimension, total linear momentum conservation furnishes one equation. Because there are two variables v_{1f} and v_{2f} to solve for in this circumstance, we require one more equation[2]. This is provided by the conservation of total kinetic energy as given by Eq. (6.11) in the case of elastic collision. Using these two equations, we obtain

$$v_{1f} = \frac{m_1 - m_2}{m_1 + m_2}v_{1i} + \frac{2m_2}{m_1 + m_2}v_{2i}, \tag{6.12}$$

$$v_{2f} = \frac{2m_1}{m_1 + m_2}v_{1i} + \frac{m_2 - m_1}{m_1 + m_2}v_{2i}. \tag{6.13}$$

When the collision is inelastic, Eq. (6.11) cannot be applied. However, for perfectly inelastic collision, we only need to solve for one variable v_f since the two particles stick together. Applying conservation of linear momentum:

$$m_1v_{1i} + m_2v_{2i} = (m_1 + m_2)v_f,$$

---

[2]In order to solve for the unknown variables, we need the same number of equations as the number of unknown variables.

we have

$$v_f = \frac{m_1 v_{1i} + m_2 v_{2i}}{m_1 + m_2} .$$

In the general case of one-dimensional inelastic collision, the additional second equation is provided by Newton's law of restitution:

$$e = \frac{v_{2f} - v_{1f}}{v_{2i} - v_{1i}} . \tag{6.14}$$

This law defines a coefficient of restitution e which gives the ratio of the relative speed after collision to that before the collision. e is an empirical quantity that depends on the nature of the material of the colliding objects. It has the range of $0 \le e \le 1$, with $e = 0$ being perfectly inelastic collision and $e = 1$ elastic collision.

When the collision is two-dimensional, four unknown variables (v_{1xf}, v_{1yf}, v_{2xf} and v_{2yf}) need to be solved. Linear momentum conservation provides two equations. But even with the additional Eq. (6.11) for elastic collision, there is a lack of one more equation. Thus for collision problems that are two-dimensions and above, additional information is required before the problem can be solved.

6.4　Elementary Physics of Many-Body Interacting Systems

6.4.1　*Center of Mass*

Up till now, we have been using a particle to represent an arbitrary object. A more useful view is that this particle representation actually corresponds to a center-of-mass representation of the object (see footnote in section 2.2). We shall clarify what this means in this section.

The center of mass of an object possesses certain characteristics. Firstly, it is a special point in a system that moves as if the total mass of the system M is concentrated at that point. Secondly, the system moves as if an external force were applied to a single particle of mass M located at the center of mass. Formally, if $\vec{r}_i$ is the position vector of a discrete particle i of mass m_i with respect to a coordinate system, the position vector of the center of mass of the system of N particles constituting the object is defined as

$$\vec{r}_{CM} = \frac{\sum_{i=1}^{N} m_i \vec{r}_i}{M} , \tag{6.15}$$

where $M = \sum_{i=1}^{N} m_i$. In terms of the Cartesian coordinate system, with $\vec{r}_{CM} = x_{CM}\hat{i} + y_{CM}\hat{j} + z_{CM}\hat{k}$ and $\vec{r}_i = x_i\hat{i} + y_i\hat{j} + z_i\hat{k}$, the center of mass of the system is defined as

$$x_{CM} = \frac{\sum_{i=1}^{N} m_i x_i}{M},$$

$$y_{CM} = \frac{\sum_{i=1}^{N} m_i y_i}{M},$$

$$z_{CM} = \frac{\sum_{i=1}^{N} m_i z_i}{M}.$$

For a system that possesses a continuous mass distribution with each mass element being Δm, the position vector of the center of mass is given by

$$\vec{r}_{CM} = \frac{1}{M} \int \vec{r} \, dm, \tag{6.16}$$

with Cartesian components

$$x_{CM} = \frac{1}{M} \int x \, dm,$$

$$y_{CM} = \frac{1}{M} \int y \, dm,$$

$$z_{CM} = \frac{1}{M} \int z \, dm.$$

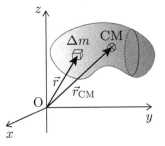

Fig. 6.1

6.4.2   *Mechanics of Many-Body Interacting System and the Conservation of Linear Momentum*

We assume a system of N discrete particles without any particle entering or leaving the system. The mass of the system M thus remains constant.

If we were to differentiate Eq. (6.15) with respect to time, we obtain the following for *the velocity of the center of mass* ($\vec{v}_{CM}$) of the system:

$$\vec{v}_{CM} = \frac{d\vec{r}_{CM}}{dt},$$

$$= \frac{1}{M} \sum_{i=1}^{N} m_i \frac{d\vec{r}_i}{dt},$$

$$= \frac{\sum_{i=1}^{N} m_i \vec{v}_i}{M}. \tag{6.17}$$

We can reexpress this equation as follows

$$M\vec{v}_{CM} = \sum_{i=1}^{N} m_i \vec{v}_i = \sum_{i=1}^{N} \vec{p}_i = \vec{p}_{tot}. \tag{6.18}$$

It states that the total linear momentum of the system equals the total mass multiplied by the velocity of the center of mass.

If we were to further differentiate Eq. (6.17) with respect to time, we obtain *the acceleration of the center of mass* ($\vec{a}_{CM}$) of the system:

$$\vec{a}_{CM} = \frac{d\vec{v}_{CM}}{dt},$$

$$= \frac{1}{M} \sum_{i=1}^{N} m_i \frac{d\vec{v}_i}{dt},$$

$$= \frac{\sum_{i=1}^{N} m_i \vec{a}_i}{M}. \tag{6.19}$$

Rearranging this equation and applying Newton's second law, we obtain

$$M\vec{a}_{CM} = \sum_{i=1}^{N} m_i \vec{a}_i = \sum_{i=1}^{N} \vec{F}_i, \tag{6.20}$$

where $\vec{F}_i$ is the net force on particle i. Note that $\sum_{i=1}^{N} \vec{F}_i$ is a sum of all the internal and external forces acting on the particles of the system, with each $\vec{F}_i$ acting at the spatial point of the ith-particle. Because all the internal forces cancel out in consequence of Newton's third law, $\sum_{i=1}^{N} \vec{F}_i = \sum \vec{F}_{ext}$. This deduction leads to the following formulation of Newton's second law for a system of particles:

$$\sum \vec{F}_{ext} = M\vec{a}_{CM}. \tag{6.21}$$

It states that the net external force on a system of particles equals the total mass of the system multiplied by the acceleration of the center of mass.

Both Eqs. (6.18) and (6.21) provide an important perspective on the mechanics of a system of interacting particles (or many-body system). They both demonstrate that the center of mass of a system of particles with combined mass M moves like an equivalent particle of mass M with instantaneous velocity $\vec{v}_{CM}$, acceleration $\vec{a}_{CM}$, and momentum $\vec{p}_{tot}$, under the influence of the net external force $\sum \vec{F}_{ext}$ acting on the system.

If the net external force $\sum \vec{F}_{ext} = 0$, Eq. (6.21) shows that

$$M\vec{a}_{CM} = M\frac{d\vec{v}_{CM}}{dt} = 0\,,$$

so that

$$M\vec{v}_{CM} = \vec{p}_{tot} = \vec{c}\,, \tag{6.22}$$

where $\vec{c}$ is a constant vector. Thus, for an isolated system of particles, the total linear momentum of the system is constant in time. This is a statement of the *Conservation of Linear Momentum* for a system of particles. In addition, we observe that the velocity of the center of mass of the system is also constant in time.

6.5 Three Conceptual Examples

6.5.1 *Impulsive Projectile Collision and the Conservation of Linear Momentum*

A projectile was launched and its trajectory follows a parabolic curve as it traverses under the action of the gravitational force (see Fig. 6.2). As it reaches the maximum height above ground, it comes into collision with an object of the same mass m which was thrown vertically upward and also reaches its maximum height at the instant it comes into collision with the projectile. During the collision, even though both projectile and object are subjected to the gravitational force, the impulsive forces $\vec{F}$ and $-\vec{F}$ that act between them are so large that we can ignore the presence of the gravitational force. This is an assumption known as the impulsive approximation. With this approximation, we can claim that the projectile and the object are not under the action of any external force, justifying the application of conservation of linear momentum to determine the eventual motions of the projectile and object after collision.

Applying conservation of linear momentum to this collision problem, we have

$$m\vec{v} = m\vec{v}_1 + m\vec{v}_2\,,$$

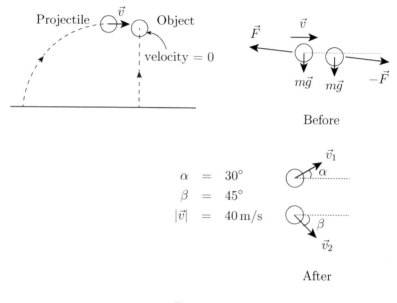

Fig. 6.2

where $\vec{v}$ is the velocity of the projectile before collision, and $\vec{v}_1$ and $\vec{v}_2$ are the respective velocities of the projectile and object after collision. Noting that $|\vec{v}| = 40$ m/s, and the direction of $\vec{v}_1$ is $\alpha = 30°$ above the horizontal while that of $\vec{v}_2$ is $\alpha = 45°$ below the horizontal after the collision as shown in Fig. 6.2, we have the above expression in component form as follows:

$$v = v_1 \cos \alpha + v_2 \cos \beta, \qquad (6.23)$$

$$0 = v_1 \sin \alpha - v_2 \sin \beta. \qquad (6.24)$$

From Eq. (6.24), we have

$$v_1 \sin 30° = v_2 \sin 45°,$$

which leads to

$$v_1 = \sqrt{2} v_2. \qquad (6.25)$$

Substituting this result into Eq. (6.23), we obtain

$$\sqrt{2} v_2 \cos 30° + v_2 \cos 45° = 40.$$

Solving for v_2 from this expression and then v_1 from Eq. (6.25), we arrive at $v_2 = 20.7$ m/s and $v_1 = 29.3$ m/s.

6.5.2 Linear Momentum and Kinetic Energy of a System with respect to Different Inertial Reference Frame

In this example, we observe the elastic collision of two identical particles A and B of mass m from the perspective of three different inertial reference frames[3]. The first reference frame is the Center-of-Mass (COM) frame[4]. The COM frame is a reference frame that moves with the COM of the system. With respect to this frame, the two particles are observed to move towards each other with speed v as shown in Fig. 6.3. Due to conservation of linear momentum and the fact that the collision is elastic, the two particles move away from each other in opposite direction with the same speed v after collision as shown in Fig. 6.3. Thus, the COM of the system is stationary at all time. Although the COM frame is in general a non-inertial reference frame, it is an inertial reference frame in this case. Note that the total linear momentum of this system of two colliding particles is zero, and its total kinetic energy is $(mv^2/2 + mv^2/2) = mv^2$. There is no change to the kinetic energy of each particle before and after the collision.

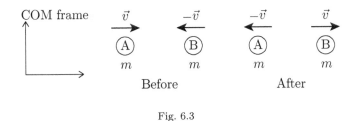

Fig. 6.3

Our second reference frame is an inertial reference frame that moves at a speed v to the right relative to the COM frame. Based on this frame, particle A is observed to be at rest while particle B is observed to move to the left with speed $2v$ before collision (see Fig. 6.4). After the collision, particle A is observed to move to the left with speed $2v$ while particle B comes to rest. The total linear momentum of this system of two particles is conserved at a value of $-2mv$, and its total kinetic energy is conserved at $m(2v)^2/2 = 2mv^2$. We notice that particle B transfers all its kinetic energy to particle A through the collision.

---

[3]To simplify discussion, we restrict the collision to 1 dimension.
[4]Note that the Center-of-Mass frame will be covered in greater detail in section 7.3.4.1.

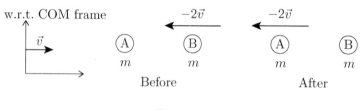

Fig. 6.4

The third scenario is with respect to an inertial reference frame that moves at a speed v to the left of the COM frame. The collision of particles A and B is now inverse to that of the second reference frame (see Fig. 6.5). The total linear momentum of the system in this case maintains constant at $2mv$ and the total kinetic energy is constant at $2mv^2$. In this case, particle A transfers all its kinetic energy to particle B via the collision.

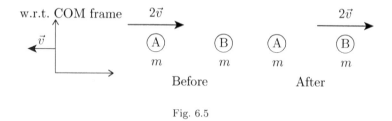

Fig. 6.5

The purpose of this example is to demonstrate that kinetic energy and linear momentum of a system are a *relative* measure that depends on the inertial reference frame. No object can be said to have a unique kinetic energy and linear momentum. Indeed, the manner in which energy and momentum is transferred is interpreted differently in all the above inertial reference frames. But it is important to note that although the kinetic energy and linear momentum differs across the different inertial reference frames, the principle of conservation of energy and conservation of momentum maintains and remains *absolute* in each case.

6.5.3 *Interesting Facts about Elastic Collision*

The example here illustrates two interesting facts concerning elastic collision. In the physical world, elastic collision happens between collision of billiard balls and in sub-atomic particle collision in a bubble chamber. *Fact 1*: if a collision is elastic in one inertial reference frame, it is elastic in all other inertial reference frames. *Fact 2:* The elastic collision between two identical particles with one initially at rest would lead typically to perpendicular trajectories between the two particles after the collision.

To prove fact 1, let us first consider an isolated system of N particles with the i-th particle having a mass of m_i. Then, the total mass of the system is $\sum_{i=1}^{N} m_i = M$. With respect to an inertial reference frame A, the velocity of the i-th particle is observed to be $\vec{v}_i$. In consequence, the total linear momentum $\vec{P}$ and total kinetic energy of this system of particles are given by the expressions:

$$\vec{P} = \sum_{i=1}^{N} m_i \vec{v}_i \,, \qquad (6.26)$$

$$T = \frac{1}{2} \sum_{i=1}^{N} m_i |\vec{v}_i|^2 \,. \qquad (6.27)$$

Let us now take reference from another inertial reference frame B which moves at a uniform velocity $\vec{u}$ relative to A. In order to determine the velocity of particle i relative to B, we employ the following relative velocity formula:

$$\vec{v}_{i/A} = \vec{v}_{i/B} + \vec{v}_{B/A} \,.$$

This formula implies that

$$\vec{v}_{i/B} = \vec{v}_{i/A} - \vec{v}_{B/A} \,,$$
$$= \vec{v}_i - \vec{u} \,, \qquad (6.28)$$

Based on this result, we can calculate the total kinetic energy of the system relative to frame B as follows:

$$T' = \frac{1}{2} \sum_{i=1}^{N} m_i |\vec{v}_i - \vec{u}|^2 \,,$$

$$= \frac{1}{2} \sum_{i=1}^{N} m_i (\vec{v}_i - \vec{u}) \cdot (\vec{v}_i - \vec{u}) \,,$$

$$= \frac{1}{2} \sum_{i=1}^{N} m_i |\vec{v}_i|^2 + \frac{1}{2} \sum_{i=1}^{N} m_i |\vec{u}|^2 - \sum_{i=1}^{N} m_i \vec{v}_i \cdot \vec{u} \,.$$

With Eq. (6.26), Eq. (6.27), $\sum_{i=1}^{N} m_i = M$ and $|\vec{u}| = u$, we can express the last line of the above equation into the following form:

$$T' = T + \frac{1}{2} M u^2 + \vec{P} \cdot \vec{u} \,. \qquad (6.29)$$

Equation (6.29) has the following implications. If the collision is elastic in frame A, T is a constant. Since $\vec{P}$ is a constant vector as a result of the conservation of linear momentum, and M and $\vec{u}$ are constants, we deduce that T' is a constant. Therefore, the collision is elastic in all inertial reference frames.

Let us now proceed to validate fact 2. In the laboratory frame, we observe a particle moving to the right with velocity $v\hat{\imath}$, while its identical particle remains at rest (see Fig. 6.6). We now shift our perspective on this collision to the COM frame in the manner described in section 6.5.2. The COM frame moves at a veloctiy $v/2$ to the right relative to the laboratory frame as shown in Fig. 6.6.

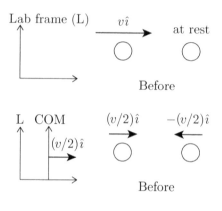

Fig. 6.6

In the COM frame, the two particles head towards each other with the same speed of $v/2$ horizontally as shown in Fig. 6.7 before the collision. Since the collision is elastic and the particles are identical, they must move away from each other with the same speed $v/2$ in opposite directions after the collision. However, the direction in which they depart from each other is arbitrary and this is captured by the unit vector $\vec{a}$ (see Fig. 6.7).

Let us now get our perspective back to the laboratory frame. For this, we need to add $v/2\,\hat{\imath}$ to each vector in the COM frame. With this addition, the velocity of particle 1 after the collision in the laboratory frame is

$$\vec{v}_1 = \frac{v}{2}\vec{a} + \frac{v}{2}\hat{\imath},$$

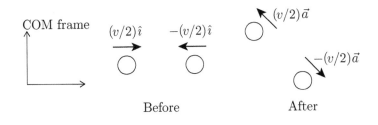

Fig. 6.7

while that of particle 2 is

$$\vec{v}_2 = -\frac{v}{2}\vec{a} + \frac{v}{2}\hat{\imath}.$$

By taking the vector dot product of these two velocities, we find that:

$$\vec{v}_1 \cdot \vec{v}_2 = \left(\frac{v}{2}\vec{a} + \frac{v}{2}\hat{\imath}\right)\left(-\frac{v}{2}\vec{a} + \frac{v}{2}\hat{\imath}\right),$$

$$= -\frac{v^2}{4}\vec{a}\cdot\vec{a} + \frac{v^2}{4}\vec{a}\cdot\hat{\imath} - \frac{v^2}{4}\hat{\imath}\cdot\vec{a} + \frac{v^2}{4}\hat{\imath}\cdot\hat{\imath},$$

$$= 0.$$

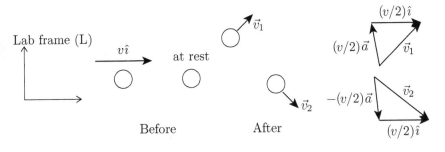

Fig. 6.8

Thus, in the laboratory frame, the two particles move apart in perpendicular directions (see Fig. 6.8). Note that this condition is true as long as $|\vec{v}_1|$ is not zero. If $|\vec{v}_1|$ is zero, or $\vec{a} = -\hat{\imath}$, particle 1 will come to rest as it transfers all its kinetic energy to particle 2, which then traverses with speed v to the right horizontally (i.e., $\vec{v}_2 = v\hat{\imath}$) in the laboratory frame (see Fig. 6.9).

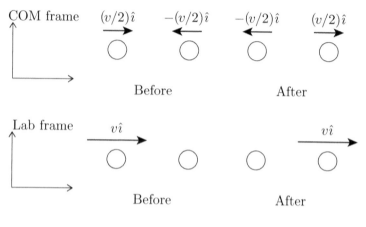

Fig. 6.9

6.6   Newton's Law with Variable Masses

Newton's second law as expressed in terms of linear momentum can be expanded mathematically as follows:

$$\vec{F} = \frac{d\vec{p}}{dt} = \frac{d\,(M\vec{v})}{dt}$$
$$= M\frac{d\vec{v}}{dt} + \vec{v}\frac{dM}{dt}\,. \tag{6.30}$$

In Chapters 3 and 4, we learnt that Newton's laws are applicable to the set of inertial reference frames. In other words, Newton's laws are supposed to be invariant with respect to the Galilean Principle of Relativity. The question now is: does the expanded form of Newton's second law as given by Eq. (6.30) continue to satisfy the Galilean Principle of Relativity?

To find out, we consider a mass M in an inertial frame S moving with velocity $\vec{v}$. The same mass when viewed in an inertial frame S' that moves with velocity $\vec{u}$ relative to S, is observed to have velocity $\vec{v}'$. Then, we expect based on Eq. (3.3) that

$$\vec{v}_{O/S} = \vec{v}_{O/S'} + \vec{v}_{S'/S}\,,$$
$$\vec{v} = \vec{v}' + \vec{u}\,.$$

If we were to substitute this result naively into Eq. (6.30), we obtain

$$\vec{F} = M \frac{d}{dt} (\vec{v}' + \vec{u}) + (\vec{v}' + \vec{u}) \frac{dM}{dt},$$

$$= M \frac{d\vec{v}'}{dt} + \vec{v}' \frac{dM}{dt} + \vec{u} \frac{dM}{dt},$$

$$= \vec{F}' + \vec{u} \frac{dM}{dt},$$ (6.31)

where $\vec{F}' = Md\vec{v}'/dt + \vec{v}'dM/dt$. Note that F is not equal to $\vec{F}'$. We notice that Eq. (6.31) is not invariant with respect to Galilean principle of relativity because of the additional term $\vec{u}\,dM/dt$. This indicates a conceptual discrepancy in the application of Newton's law of motion to the variable mass system.

We address this inconsistency by considering the following scenario, which is illustrated in Fig. 6.10. Here, an infinitesimal mass Δm with initial velocity $\vec{v}_m$ aggregates onto a mass M that moves with velocity $\vec{v}$ at $t = 0$. An external force $\vec{F}$ acts on the system such that at $t = \Delta t$, the combined mass $M + \Delta m$ continues to maintain the velocity $\vec{v}$ of M. The total linear momentum of the system at $t = 0$ is

$$\vec{P}(0) = \Delta m \vec{v}_m + M\vec{v},$$

while the total linear momentum of the system at $t = \Delta t$ is

$$\vec{P}(\Delta t) = (M + \Delta m)\,\vec{v}.$$

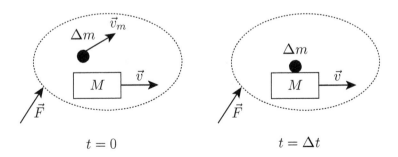

$$t = 0 \qquad\qquad t = \Delta t$$

Fig. 6.10

By Newton's second law, we have

$$\begin{aligned}
\vec{F} &= \lim_{\Delta t \to 0} \frac{\vec{P}(\Delta t) - \vec{P}(0)}{\Delta t}, \\
&= \lim_{\Delta t \to 0} \frac{(M + \Delta m)\,\vec{v} - (M\vec{v} + \Delta m\vec{v}_m)}{\Delta t}, \\
&= \lim_{\Delta t \to 0} \left(\vec{v}\frac{\Delta m}{\Delta t} - \vec{v}_m \frac{\Delta m}{\Delta t} \right), \\
&= (\vec{v} - \vec{v}_m) \frac{dm}{dt}.
\end{aligned}$$

Noting that $\Delta m = \Delta M$, we have

$$\vec{F} = (\vec{v} - \vec{v}_m)\frac{dM}{dt}. \tag{6.32}$$

Comparing this result to Eq. (6.30) where $\vec{F} = \vec{v}dM/dt$ when $d\vec{v}/dt = 0$, we observe an additional term $\vec{v}_m dM/dt$ in Eq. (6.32) due to the initial velocity of Δm. This additional term is in fact important in ensuring that Newton's law is consistent with Galilean principle of relativity. We can verify this fact by taking the perspective of the inertial reference frame S' where $\vec{v}' = \vec{v} - \vec{u}$ and $\vec{v}_m{}' = \vec{v}_m - \vec{u}$. Then,

$$\begin{aligned}
\vec{F}' &= (\vec{v}' - \vec{v}_m')\frac{dM}{dt}, \\
&= (\vec{v} - \vec{u} - \vec{v}_m + \vec{u})\frac{dM}{dt}, \\
&= (\vec{v} - \vec{v}_m)\frac{dM}{dt}, \\
&= \vec{F}. \tag{6.33}
\end{aligned}$$

Thus, Eq. (6.32) is invariant with respect to Galilean transformation.

We can in fact extend the above scenario to the more general case where the external force $\vec{F}$ causes the combined mass to attain a velocity $\vec{v} + \Delta\vec{v}$ (instead of just maintaining it at $\vec{v}$) at $t + \Delta t$ (see Fig. 6.11). In this case, we have

$$\begin{aligned}
\vec{P}(0) &= \Delta m\vec{v}_m + M\vec{v}, \\
\vec{P}(\Delta t) &= (M + \Delta m)(\vec{v} + \Delta\vec{v}).
\end{aligned}$$

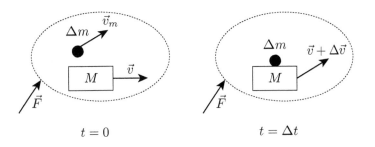

Fig. 6.11

Again, by Newton's second law, we have

$$
\begin{aligned}
\vec{F} &= \lim_{\Delta t \to 0} \frac{\vec{P}(\Delta t) - \vec{P}(0)}{\Delta t}, \\
&= \lim_{\Delta t \to 0} \frac{M\vec{v} + M\Delta\vec{v} + \Delta m\vec{v} + \Delta m\Delta\vec{v} - M\vec{v} - \Delta m\vec{v}_m}{\Delta t}, \\
&= \lim_{\Delta t \to 0} \left(M\frac{\Delta\vec{v}}{\Delta t} + (\vec{v} - \vec{v}_m)\frac{\Delta m}{\Delta t} + \Delta m\frac{\Delta\vec{v}}{\Delta t} \right), \\
&= M\frac{d\vec{v}}{dt} + (\vec{v} - \vec{v}_m)\frac{dm}{dt}.
\end{aligned}
$$

Because $\Delta m = \Delta M$, we obtain

$$
\vec{F} = M\frac{d\vec{v}}{dt} + (\vec{v} - \vec{v}_m)\frac{dM}{dt}. \tag{6.34}
$$

Notice the difference between Eq. (6.34) and Eq. (6.30), and more importantly, Eq. (6.34) obeys Galilean principle of relativity while Eq. (6.30) does not. The physical meaning of Eq. (6.30) will be discussed at the end of the section.

In the previous two scenarios, we have considered mass aggregation. Let us now consider the case of mass ejection (see Fig. 6.12). At $t = 0$, the mass $M + \Delta m$ moves with velocity $\vec{v}$ and is acted on by a force $\vec{F}$. At $t = \Delta t$, an infinitesimal mass Δm is ejected with velocity $\vec{v}_m$, while the mass M attains a velocity $\vec{v} + \Delta\vec{v}$, as the entire system continues to be acted on by $\vec{F}$. Under this circumstance, we have

$$
\vec{P}(0) = (M + \Delta m)\,\vec{v},
$$
$$
\vec{P}(\Delta t) = \Delta m\vec{v}_m + M\,(\vec{v} + \Delta\vec{v}).
$$

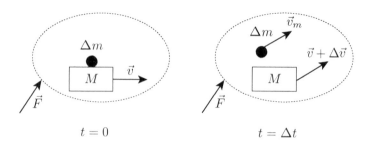

$$t = 0 \qquad\qquad\qquad t = \Delta t$$

Fig. 6.12

Applying Newton's second law, we obtain

$$\vec{F} = \lim_{\Delta t \to 0} \frac{\vec{P}(\Delta t) - \vec{P}(0)}{\Delta t},$$

$$= \lim_{\Delta t \to 0} \frac{\Delta m\,\vec{v}_m + M\vec{v} + M\Delta\vec{v} - M\vec{v} - \Delta m\,\vec{v}}{\Delta t},$$

$$= \lim_{\Delta t \to 0} \left(M\frac{\Delta\vec{v}}{\Delta t} + (\vec{v}_m - \vec{v})\frac{\Delta m}{\Delta t} \right),$$

which gives

$$\vec{F} = M\frac{d\vec{v}}{dt} + (\vec{v}_m - \vec{v})\frac{dm}{dt}. \tag{6.35}$$

Since $\Delta M = -\Delta m$, we recover Eq. (6.34) from Eq. (6.35) for mass ejection.

6.6.1　*Application to Variable Mass Systems*

Let us now apply our formulation of Newton's law for variable masses to the problem of rocket propulsion and sand falling onto a freight car.

6.6.1.1　*Rocket Propulsion*

In the reference frame of the rocket, its mass M and an infinitesimal element of its fuel Δm are both stationary at $t = 0$. At $t = \Delta t$, the fuel Δm is ejected backward with velocity $\vec{v}_e = -v_e\hat{\imath}$, while the rocket propels forward with velocity $\Delta\vec{v} = \Delta v\hat{\imath}$. In the reference frame of Earth, the rocket and the fuel element both travels at velocity $\vec{v} = v\hat{\imath}$ at $t = 0$. At $t = \Delta t$, the fuel element is observed with velocity $\vec{v}_m = (v - v_e)\,\hat{\imath}$, while the rocket with velocity $\vec{v} + \Delta\vec{v} = (v + \Delta v)\,\hat{\imath}$. Thus, the scenario of the fuel element Δm being ejected from the rocket from the reference frame of the Earth, is exactly the same as the formulation of Newton's law for mass ejection above. Applying Eq. (6.34) to this context, we have

$$M\frac{d\vec{v}}{dt} + (\vec{v} - \vec{v}_m)\frac{dM}{dt} = 0,$$

$$M\frac{dv}{dt}\hat{\imath} + (v\hat{\imath} - v\hat{\imath} + v_e\hat{\imath})\frac{dM}{dt} = 0,$$

where we have assumed that the external force (such as gravity) is negligible compared to the internal forces that act between the rocket and the fuel. Upon simplifying the above results, we obtain

$$\text{Thrust} = M\frac{dv}{dt} = -v_e\frac{dM}{dt}, \tag{6.36}$$

where $M\,dv/dt$ is due to the reaction force from the fuel gas on the rocket. Hence, $M\,dv/dt$ is also known as the thrust.

6.6.1.2 *Sand Falling onto Freight Car*

This example involves sand falling from a stationary hopper onto a freight car which is moving with uniform velocity $\vec{v}$ in the horizontal direction. The sand falls at the rate dm/dt How much horizontal force $\vec{F}$ is needed to keep the freight car moving at the same velocity?

With the freight car M moving with constant velocity $\vec{v}$, $d\vec{v}/dt = 0$. Equation (6.34) then gives

$$\vec{F} = (\vec{v} - \vec{v}_m)\frac{dM}{dt}.$$

Note that $\vec{v} = v\hat{\imath}$ and $\vec{v}_m = v_m\hat{\jmath}$. Since the applied force $\vec{F} = F\hat{\imath}$ is necessarily in the horizontal direction, we shall analyze the above equation in the x-direction. With $dM/dt = dm/dt$, we obtain

$$F = v\frac{dm}{dt}. \tag{6.37}$$

Finally, let us consider what does Eq. (6.30) mean physically. First, note that the term $M d\vec{v}/dt$ has the connotation that a force is required to change the velocity of mass M. Second, this gives the physical context that a force is required to change the *intrinsic* mass M through the term $\vec{v}dM/dt$,

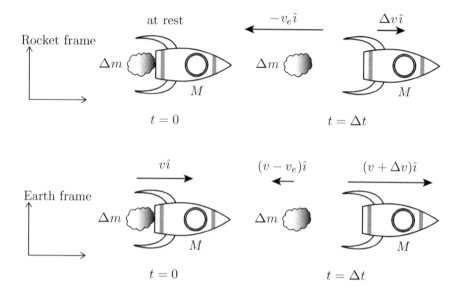

Fig. 6.13

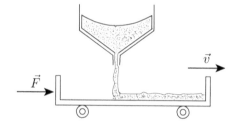

Fig. 6.14

which is different from the *extrinsic* mass change expressed through Eqs. (6.32) and (6.34)[5].

In fact, such an intrinsic change in mass occurs in special relativity, when the work done by a force as formulated in Eq. (6.30) leads to an increase in relativistic energy, or in other words, *intrinsic* mass. We illustrate how

---

[5]Here, extrinsic mass change refers to the injection and ejection of mass into the system at non-relativistic speed, while intrinsic mass change results from work done by forces on an object near the speed of light.

this come about by first calculating the kinetic energy K:

$$K = \int_{u'=0}^{u'=u} F dx,$$

$$= \int_{u'=0}^{u'=u} \frac{d}{dt}(m'u')dx,$$

$$= \int_{u'=0}^{u'=u} d(m'u')\frac{dx}{dt},$$

$$= \int_{u'=0}^{u'=u} (m'du' + u'dm')u',$$

$$= \int_{u'=0}^{u'=u} (m'u'du' + u'^2 dm').$$

Notice that Eq. (6.30) appears in the fourth line of the above equation. By using $m'^2c^2 - m'^2u'^2 = m_0^2 c^2$, and taking differentials in this equation through division by $2m'$, we have

$$2m'c^2 dm' - 2m'^2 u' du' - 2u'^2 m' dm' = 0,$$

$$m'u'du' + u'^2 dm' = c^2 dm'.$$

Thus,

$$K = \int_{u'=0}^{u'=u} c^2 dm' = c^2 \int_{m'=m_0}^{m'=m} dm' = mc^2 - m_0 c^2.$$

Thus, we have

$$E = mc^2 = m_0 c^2 + K. \tag{6.38}$$

Since such an "intrinsic" change in mass is not possible at speeds far below the speed of light, the form of Eqs. (6.32) and (6.34) are to be adopted in describing Newton's second law for variable mass systems in the everyday scenario of slow speeds.

As a final note, both Newton's laws and Einstein's theory of special relativity require the mechanical laws (Newton's) and the physical law (Einstein's)[6] to be invariant with respect to different inertial reference frame. This fact remains true in the above discussion. The important point to note is that while Newton's laws remain invariant with respect to Galilean transformation, Einstein's theory of special relativity remains invariant with respect to the Lorentz transformation.

---

[6]Physical laws include the mechanical laws of Newton and also Maxwell's theory of electromagnetism.

Chapter 7

The Fundamental Mechanics of Rotational Motion

7.1 Introduction

In general, motion can be classified as being either in the translational or rotational form, although it is typically a combination of both during an actual motion. In previous chapters, we have covered in detail the fundamentals of linear translational motion. While we have also touched on rudiments of rotational motion in the form of circular motion, it was done by representing an object as a particle. In this chapter, we shall go into the details of rotational motion by taking into account the spatial extent of the object. At the end, we will provide a general formulation to analyze and depict the mechanics of translational and rotational motion that occur at the same time. In particular, this chapter deals with the special case where the direction of the axis of rotation remains unchanged.

7.2 Rigid Body and Rotational Kinematics

7.2.1 *Rigid Body*

When an extended object rotates, its motion cannot be analyzed by treating it as a particle. This is because at any given point in time, different parts of the object possess different linear velocities and linear accelerations. As a result, we have to analyze the motion of each part of the object separately.

However, the analysis is greatly simplified if the rotating object is assumed to be rigid. A rigid object is not elastic. It is not deformable, like a spring with an infinite spring constant. Because of this, the relative positions of all particles within a rigid body is fixed. This condition will not change no matter how strong the force is that acts on the body. It is important to note that all real objects are not rigid. They are elastic and deformable to some extent. The assumption that a body is rigid is an approximation, and is useful for situations when the deformation of the object is negligible.

7.2.2   *Rotational Kinematics*

In this section, we shall show how the assumption of a rigid body leads to the definition of physical quantities such as the angular position, angular displacement, angular velocity, and angular acceleration of a spatially extended object.

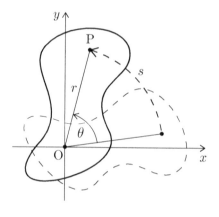

Fig. 7.1

Let us now consider the rotation of a rigid object about a fixed axis of rotation as shown in Fig. 7.1. We first identify a particle at point P of the object. This particle is located at a distance r from the origin O of the object (see Fig. 7.1). As the object rotates counter-clockwise about its axis of rotation, the position of P changes as it maintains a distance r from the origin while the line OP sweeps through an angle θ. Hence, for a rigid object rotating about a fixed axis, it is convenient to represent its point by the coordinate (r, θ), where r is the distance of the point to the fixed axis, while θ is the angle measured counter-clockwise from some fixed reference line.

Due to the fact that the object is a rigid body, when a particle moves along a circular path away from the fixed reference line through angle θ, every other particle on the object rotates through the same angle. This allows us to associate the angle θ with the entire rigid object and also to each individual particle of the object. This means that the concept of an angular position, as defined by θ, makes sense for a rigid object. Now, the angle θ has the unit of radian and is related to the arc-length s in the following way:

$$\theta = \frac{s}{r}\,. \tag{7.1}$$

More formally, the *angular position* of the rigid object is the angle between the line from the origin to an arbitrary reference point on the object (such as our line OP) and a fixed reference line in space, which is usually chosen as the x-axis. The angular position θ can be positive or negative[1]. If we choose positive angles to be measured counter-clockwise from the positive x-axis, then θ is positive. On the other hand, angles measured clockwise from the positive x-axis will be negative.

With the clear specification of angular position, we are now ready to introduce *angular displacement* of a particle, which is defined as the change in angular position of the particle within some time interval, i.e.,

$$\Delta\theta = \theta_f - \theta_i\,. \tag{7.2}$$

For example, as a rigid object rotates from angular position A to angular position B in the time interval Δt, the radius vector moves through an angular displacement $\Delta\theta = \theta_f - \theta_i$.

Note that angular displacements do not add as vector quantities for finite rotations. This can be easily understood by taking the example of first rotating an object by $\pi/2$ rad in the clockwise direction about the x-axis, followed by a $\pi/2$ rad clockwise rotation about the y-axis. If we were to put the object in the same starting condition and change the order of the operation, i.e., first a $\pi/2$ rad clockwise rotation about the y-axis, followed by a $\pi/2$ rad clockwise rotation about x-axis, we will observe a different orientation of the final rotated object in the two cases. This example demonstrates that angular displacements with finite rotations do not follow the commutative law of vector addition, and hence, such angular displacements are not vector quantities.

Next, we can define *angular velocity* of the object as the rate of change of angular displacement. There are two perspectives on angular velocity. The first perspective is the *average angular velocity* $\bar{\omega}$, which is defined

---

[1]This is similar to motion in a straight line where the specification of the direction of positive displacement is necessary.

as the ratio of the angular displacement $\Delta\theta$ of a rigid object to the time interval Δt during which the displacement occurs:

$$\bar{\omega} = \frac{\theta_f - \theta_i}{t_f - t_i} = \frac{\Delta\theta}{\Delta t}. \tag{7.3}$$

The second perspective is the *instantaneous angular velocity* ω, which is defined as the limit of the average angular velocity as Δt approaches zero, i.e.,

$$\omega = \lim_{\Delta t \to 0} \frac{\Delta\theta}{\Delta t} = \frac{d\theta}{dt}. \tag{7.4}$$

The unit of both version of angular velocities are rad/s.

Since the average angular velocity depends on the angular displacement $\Delta\theta$, it is not a vector quantity. However, an infinitesimal version of $\Delta\theta$ (i.e., an infinitesimally small rotation) can be shown to obey the properties of vector. This implies that instantaneous angular velocity is a vector quantity. Thus, we can express the instantaneous angular velocity as $\vec{\omega}$. Its direction is given by the right-hand rule: when four fingers of the right hand wrap in the direction of rotation, the direction of the extended right thumb gives the direction of $\vec{\omega}$. Based on this convention, the direction of $\vec{\omega}$ for a rigid object that rotates about a fixed axis is in the direction of the fixed axis. Under this circumstance, we do not need to employ vector notation during analysis (just like one-dimension translational kinematics). The vectorial nature of $\vec{\omega}$ is already implicit in the sign of ω: positive for counter-clockwise rotation (θ increasing), and negative for clockwise rotation (θ decreasing).

Another important rotational kinematic quantity is the *angular acceleration*. It is defined as the rate of change of angular velocity. Again, there are two perspectives on angular acceleration: average angular acceleration and instantaneous angular acceleration. *Average angular acceleration* $\bar{\alpha}$ of a rotating rigid object is defined as the ratio of the change in the angular velocity to the time interval Δt during which the change in the angular velocity occurs:

$$\bar{\alpha} = \frac{\omega_f - \omega_i}{t_f - t_i} = \frac{\Delta\omega}{\Delta t}. \tag{7.5}$$

Instantaneous angular acceleration α is defined as the limit of the average angular acceleration as Δt approaches zero:

$$\alpha = \lim_{\Delta t \to 0} \frac{\Delta\omega}{\Delta t} = \frac{d\omega}{dt}. \tag{7.6}$$

The unit of angular acceleration is $\mathrm{rad/s^2}$. Just like instantaneous angular velocity, instantaneous angular acceleration is a vector quantity. In fact, $\vec{\alpha} = d\vec{\omega}/dt$. For rotation about a fixed axis, the direction of $\vec{\alpha}$ is along the fixed axis because the direction of $\vec{\omega}$ always lies along this axis. In this case, $\vec{\alpha}$ and $\vec{\omega}$ is parallel to each other if the angular velocity is increasing in time, and they are anti-parallel to each other if the angular velocity is decreasing in time. Finally, based on the same reasoning as above, the average angular acceleration is not a vector quantity.

During the rotation of a rigid body, every particle on the object rotates through the same angle in a given time interval and has the same angular velocity and the same angular acceleration. Hence, the kinematic quantities θ, $\vec{\omega}$, and $\vec{\alpha}$ as defined in this section characterize the rotational motion of the entire rigid body as well as individual particles in the object.

For rotation about a fixed axis, these rotational kinematic quantities can all be expressed without vector notation, i.e., θ, ω, and α. They in fact correspond to the linear kinematic quantities in the following fashion: angular position (θ) to linear position (x); angular velocity (ω) to linear velocity (v); and angular acceleration (α) to linear acceleration (a). In fact, the variables θ, ω, and α differ from x, v, and a dimensionally only by a factor having the unit of length which we shall show later. This correspondence between rotational and translational kinematic quantities is especially clear when the angular acceleration α is a constant. Under this condition, we have the following set of rotational kinematic equations:

$$\omega_f = \omega_i + \alpha t \, , \tag{7.7}$$

$$\theta_f = \theta_i + \omega_i t + \frac{1}{2}\alpha t^2 \, , \tag{7.8}$$

$$\omega_f^2 = \omega_i^2 + 2\alpha \left(\theta_f - \theta_i\right) \, , \tag{7.9}$$

$$\theta_f = \theta_i + \frac{1}{2}\left(\omega_i + \omega_f\right) t \, , \tag{7.10}$$

which corresponds to the set of translational kinematic equations in one-dimension:

$$v_f = v_i + at \, , \tag{7.11}$$

$$x_f = x_i + v_i t + \frac{1}{2}at^2 \, , \tag{7.12}$$

$$v_f^2 = v_i^2 + 2a\left(x_f - x_i\right) \, , \tag{7.13}$$

$$x_f = x_i + \frac{1}{2}\left(v_i + v_f\right) t \, . \tag{7.14}$$

Angular and linear quantities are in fact closely related. This can be observed by first expressing the linear velocity of a particle in terms of polar coordinate as follows:

$$\vec{v} = v_r \hat{r} + v_t \hat{\theta} \,. \tag{7.15}$$

For a rotating rigid body that moves in a circle about a fixed axis of rotation, all its particles have radial velocity $v_r = 0$. Hence, the linear velocity

$$\vec{v} = v_t \hat{\theta} = v \hat{\theta} \tag{7.16}$$

is always tangential to the circular path. In fact, the magnitude of the linear velocity $|\vec{v}| = v$ obeys

$$v = r\omega \,,$$

according to Eq. (2.46). Thus, not every point has the same v although every point on the rigid body has the same angular velocity ω. As one moves outward from the center of rotation, v increases. This shows that every particle on the rigid body rotating about a fixed axis has the same angular motion but not the same linear motion.

Similarly, for acceleration, we have

$$\vec{a} = a_t \hat{\theta} + a_r \hat{r}$$

from Eq. (2.50). Noting that $a_t = dv/dt = d(r\omega)/dt = rd\omega/dt$, we have, for the tangential acceleration

$$a_t = r\alpha \,. \tag{7.17}$$

Also, with $a_r = -v^2/r = -(r\omega)^2/r$, we obtain, for the radial acceleration

$$a_r = -r\omega^2 \,, \tag{7.18}$$

with the negative sign indicating that the radial acceleration is directed towards the fixed axis of rotation. Together, we have

$$\vec{a} = r\alpha \hat{\theta} - r\omega^2 \hat{r} \,. \tag{7.19}$$

Note that $|\vec{a}| = \sqrt{a_t^2 + a_r^2} = r\sqrt{\alpha^2 + \omega^4}$. Thus, even though all points on the rigid body have the same angular acceleration, they do not have the same tangential and radial acceleration.

7.3 Rotational Dynamics

Just like translational acceleration, there is a *cause* to rotational acceleration. That cause is a turning force known as the torque. In this section, we shall understand the effects of torque on the rotational dynamics of a rigid body. Moreover, we will expand on how force and torque act together on a rigid body to generate both its translational and rotational accelerations.

7.3.1 Torque and Angular Momentum

The torque $\vec{\tau}$ that acts on an object is defined as

$$\vec{\tau} = \vec{r} \times \vec{F}, \tag{7.20}$$

where $\vec{r}$ is the position vector from an arbitrary reference point O to the point of application of the force $\vec{F}$ on the body. $\vec{\tau}$ lies in a direction perpendicular to the plane formed by $\vec{r}$ and $\vec{F}$. Its unit is N.m.

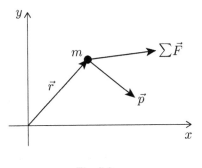

Fig. 7.2

Let us first consider from the point of view of a particle of mass m located at a point with vector position $\vec{r}$ and move with linear momentum $\vec{p}$ as shown in Fig. 7.2. If this particle is acted on by a net force $\sum \vec{F}$, we can write

$$\vec{r} \times \sum \vec{F} = \sum \vec{\tau} = \vec{r} \times \frac{d\vec{p}}{dt}. \tag{7.21}$$

We now add to the right-hand-side of this equation the term $d\vec{r}/dt \times \vec{p}$, which is zero because $d\vec{r}/dt = \vec{v}$, and $\vec{v}$ and $\vec{p}$ are parallel. Thus,

$$\sum \vec{\tau} = \vec{r} \times \frac{d\vec{p}}{dt} + \frac{d\vec{r}}{dt} \times \vec{p} = \frac{d(\vec{r} \times \vec{p})}{dt}. \tag{7.22}$$

Interestingly, this equation is similar in form to $\sum \vec{F} = d\vec{p}/dt$, which suggests that $\vec{r} \times \vec{p}$ plays the same role in rotational motion as $\vec{p}$ plays in translational motion. This leads us to the definition of *angular momentum* $\vec{L}$ of a particle relative to an arbitrary reference point O as the cross product of the particle's position vector $\vec{r}$ and its linear momentum $\vec{p}$:

$$\vec{L} = \vec{r} \times \vec{p}. \tag{7.23}$$

The unit of angular momentum is $\mathrm{kg\,m^2/s}$. Note that both the magnitude and direction of $\vec{L}$ depends on the choice of the point O. The magnitude of $\vec{L}$ is given by $mvr\sin\phi$, where ϕ is the angle between $\vec{p}$ and $\vec{r}$. In addition, the direction of $\vec{L}$ is perpendicular to the plane formed by $\vec{r}$ and $\vec{p}$ according to the right-hand rule.

Putting the definition of angular momentum into Eq. (7.22), we obtain

$$\sum \vec{\tau} = \frac{d\vec{L}}{dt}. \tag{7.24}$$

Equation (7.24) states that the torque $\sum \vec{\tau}$ acting on a particle is equal to the time rate of change of the particle's angular momentum. This equation is the rotational analog of Newton's second law. The torque causes the angular momentum $\vec{L}$ to change just as the force causes the linear momentum $\vec{p}$ to change. It is important to note that Eq. (7.24) is valid only if $\sum \tau$ and $\vec{L}$ are measured about the same reference point O. Furthermore, the expression is valid for any O that is fixed in an inertial reference frame.

7.3.2   Conservation of Angular Momentum

Let us now extend our consideration to the angular momentum of a system of particles. The total angular momentum of a system of particles is defined as the vector sum of the angular momenta of the individual particles:

$$\vec{L}_{tot} = \vec{L}_1 + \vec{L}_2 + \cdots + \vec{L}_n = \sum_{i=1}^{n} \vec{L}_i. \tag{7.25}$$

Differentiating with respect to time and then using Eq. (7.24):

$$\frac{d\vec{L}_{tot}}{dt} = \sum_{i=1}^{n} \frac{d\vec{L}_i}{dt} = \sum_{i=1}^{n} \vec{\tau}_i. \tag{7.26}$$

According to Newton's third law, any torque associated with the internal forces acting on a system of particles cancels each other, which implies that the net internal torque vanishes. Therefore, we have the important result:

The net external torque acting on a system about an arbitrary reference point O in an inertial frame equals the time rate of change of the total angular momentum of the system about that same point O:

$$\sum \vec{\tau}_{ext} = \frac{d\vec{L}_{tot}}{dt}. \tag{7.27}$$

If there is no external torque, Eq. (7.27) implies that the total angular momentum is conserved. In other words,

The total angular momentum of a system is constant in both magnitude and direction if the resultant external torque acting on the system is zero, i.e., if the system is isolated:

$$\vec{L}_{tot} = \sum_{i=1}^{n} \vec{L}_i = \vec{c},\tag{7.28}$$

where $\vec{c}$ is a constant vector in time, or

$$\vec{L}_{initial} = \vec{L}_{final}.\tag{7.29}$$

Finally, while the above results are formulated for a system of particles, they are directly applicable to a rigid object since such an object is a special case of a system of many particles.

7.3.3 Rotation about a Fixed Axis

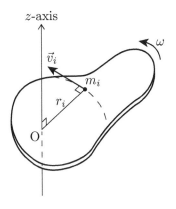

Fig. 7.3

In this section, we shall consider a special case of general rotation of a rigid object, where rotation is constrained about a fixed axis. Let us assume that this fixed axis coincides with the z-axis of the coordinate system. Then, each particle of the rigid object shall rotate in a circular path in the x-y plane about the z-axis with an angular speed of ω (see Fig. 7.3). The angular momentum $\vec{L}_i$ of each individual particle i of this object is given by

$$\vec{L}_i = \vec{r}_i \times \vec{p}_i \,,$$
$$= \vec{r}_i \times m_i \vec{v}_i \,,$$
$$= m_i r_i v_i \hat{k} \,,$$
$$= m_i r_i^2 \omega \hat{k} \,, \tag{7.30}$$

where the velocity of the particle $\vec{v}_i$ is perpendicular to its position vector $\vec{r}_i$ (which is taken with respect to the fixed axis) at all times. Because $\vec{r}_i$ and $\vec{v}_i$ lies in the x-y plane, their cross product gives the direction of $\vec{L}_i$ (and also $\vec{\omega}$) which is in the direction of the z axis. In other words, for rotation about a fixed axis, the reference point O is always chosen on the fixed axis for each respective horizontal slice (which is parallel to the x-y plane) that constitutes the structure of the whole three-dimensional rigid body (see Fig. 7.4).

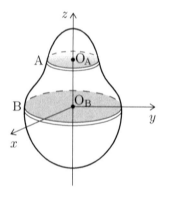

Fig. 7.4

By summing $\vec{L}_i$ over all particles, we obtain the total angular momentum of the whole object:

$$\vec{L} = \sum_{i=1}^{n} \vec{L}_i \,,$$
$$= \sum_{i=1}^{n} m_i r_i^2 \omega \hat{k} \,,$$
$$= \left(\sum_{i=1}^{n} m_i r_i^2 \right) \omega \hat{k} \,,$$
$$= I \omega \hat{k} \,, \tag{7.31}$$

where we have defined

$$I = \sum_{i=1}^{n} m_i r_i^2 \qquad (7.32)$$

and termed it as the *moment of inertia*. The moment of inertia is a measure of the resistance of an object to changes in rotational motion, just as mass is a measure of the tendency of an object to resist changes in linear motion. It has a unit of kg m^2. The moment of inertia of an extended rigid object can be obtained by dividing the object into many small volume elements, each of which has mass Δm_i. By using the definition of I given in Eq. (7.32) and taking the limit of this sum as $\Delta m_i \to 0$, we obtain the following expression of the moment of inertia of a rigid body

$$I \equiv \lim_{\Delta m_i \to 0} \sum_i r_i^2 \Delta m_i = \int r^2 dm . \qquad (7.33)$$

Note that I depends on the axis of rotation, and upon the size and shape of the object.

Equation (7.31) informs us that

$$\vec{L} = L_z \hat{k} = I \omega \hat{k} . \qquad (7.34)$$

Differentiating this equation with respect to time, we have

$$\frac{d\vec{L}}{dt} = \frac{dL_z}{dt} \hat{k} = I \frac{d\omega}{dt} \hat{k} = I \alpha \hat{k} = I \vec{\alpha} ,$$

where $\vec{\alpha} = \alpha \hat{k} = d\omega/dt \, \hat{k}$ is the angular acceleration relative to the axis of rotation. As $d\vec{L}/dt = dL_z/dt \, \hat{k}$ is equal to the external torque (see Eq. (7.27)), we have

$$\sum \vec{\tau}_{ext} = I \alpha \hat{k} . \qquad (7.35)$$

Note that Eq. (7.35) is also valid for a rigid object rotating about a moving axis provided the moving axis passes through the center of mass and also is an axis of symmetry. Moreover, we should note that in general, the individual external torque need not point along the z-axis[2]. It is the net external torque that is in the direction of the z-axis. Components of the individual external torque that do not lie in the z-axis direction have been canceled by external torque from the constrained (or reaction) forces due to the fixed axis.

---

[2]In other words, the individual external force that acts on the body need not lie in the x-y plane.

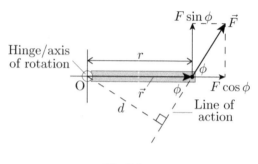

Fig. 7.5

According to Eq. (7.20), for a torque $\vec{\tau}$ that lies in the direction of the z-axis which arises from a force $\vec{F}$ that acts in the x-y plane (see Fig. 7.5), we can express this torque explicitly as

$$\vec{\tau} = \vec{r} \times \vec{F} = \tau \hat{k} = rF \sin\phi\,\hat{k} = Fd\,\hat{k}\,, \qquad (7.36)$$

where F is the magnitude of the applied force $\vec{F}$, ϕ is the angle between F and the line drawn from the reference point O at the fixed axis to the point of application of F, and r is the distance of this line (see Fig. 7.5). Note that d is the perpendicular distance from the point O to the line of action of $\vec{F}$, and is known as the *moment arm*. Conventionally, this perspective provides the notion that torque is a turning force, where the component $F \sin\phi$ of $\vec{F}$ multiplied by d is known as the moment of the force and contributes to the rotation of the object. $F \sin\phi$ is also known as the tangential force and the torque generated by it is related to the angular acceleration α of the object according to Eq. (7.35) as follows:

$$Fd = I\alpha\,.$$

On the other hand, the component $F \cos\phi$ of $\vec{F}$ has no tendency to produce rotation based on the manner in which it acts on the object. $F \cos\phi$ is also known as the radial force. Such a radial component of the force produces zero torque because the line of action of all radial components must pass through the axis of rotation. The vector nature of torque is clearly exhibited by whether $\vec{\tau}$ points in the direction of $+z$ or $-z$. In the former case $(0 < \phi < \pi)$, the component of the torque is positive and the nature of the force is to turn the object counter-clockwise. For the latter case $(\pi < \phi < 2\pi)$, the component of torque is negative as the turning tendency is clockwise. Thus, force not only causes a change in linear motion, it can also cause a change in rotational motion. The effectiveness of this change in rotational motion depends on the force and the moment arm, which is quantified by torque.

In summary, we can write Eqs. (7.34) and (7.35) in component form as follows:

$$L_z = I\omega, \qquad (7.37)$$

$$\tau_{ext} = I\alpha, \qquad (7.38)$$

and drop the vector notation. Such a formulation of rotational quantities is sufficient for the treatment of the rotational dynamics of a rigid body about a fixed axis.

In the last section, we learnt that in the absence of external torque, the principle of conservation of angular momentum applies as the system is isolated. Now, if the mass of such a system were to undergo redistribution, we will expect the moment of inertia to change. The conservation of angular momentum then requires a compensating change in the angular velocity in the following manner:

$$L_{tot} = I_i\omega_i = I_f\omega_f = \text{constant}. \qquad (7.39)$$

Equation (7.39) is valid for rotation about a fixed axis and for rotation about an axis through the center of mass of a moving system as long as that axis remains fixed in direction.

Let us now determine the energy of a rigid body rotating about a fixed axis. Such an object has no translational motion but only rotational motion, and we expect the presence of kinetic energy in the rotational form. For this, we first consider a single particle (i.e., the i-th particle) on the body. The kinetic energy of K_i of this particle is given by its mass m_i and tangential velocity $\vec{v}_i$:

$$K_i = \frac{1}{2}m_i v_i^2 = \frac{1}{2}m_i r_i^2 \omega^2, \qquad (7.40)$$

where $v_i = |\vec{v}_i| = r_i\omega$. Note that every particle in the object has the same angular velocity ω. The total kinetic energy of the rotating rigid object is the sum of the kinetic energies of the individual particles:

$$K_R = \sum_i K_i = \sum_i \frac{1}{2}m_i v_i^2 = \frac{1}{2}\left(\sum_i m_i r_i^2\right)\omega^2. \qquad (7.41)$$

Thus, the total rotational kinetic energy of the rotating rigid object is

$$K_R = \frac{1}{2}I\omega^2, \qquad (7.42)$$

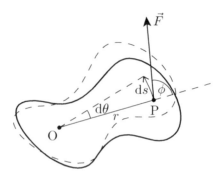

Fig. 7.6

where I is the moment of inertia. From Eq. (7.42), we see that the rotational kinetic energy $I\omega^2/2$ is analogous to the translational kinetic energy $mv^2/2$. The quantities I and ω in rotational motion directly correspond to m and v in linear motion, respectively.

Now, suppose a force $\vec{F}$ acts on a rigid body as shown in Fig. 7.6. The body rotates through an infinitesimal angle $d\theta$ about a fixed axis during an infinitesimal time interval dt. The work dW done by $\vec{F}$ while the point P moves a distance ds is $dW = F\sin\phi\, ds$. Since $ds = r\,d\theta$, we have $dW = Fr\sin\phi\, d\theta$, or from Eq. (7.36),

$$dW = \tau d\theta. \tag{7.43}$$

The total work W done by the torque during an angular displacement from θ_i to θ_f is

$$W = \int_{\theta_i}^{\theta_f} \tau d\theta. \tag{7.44}$$

If the torque is constant and $\Delta\theta = \theta_f - \theta_i$, then $W = \tau(\theta_f - \theta_i) = \tau\Delta\theta$.

The rate at which work is being done by $\vec{F}$ as the object rotates about the fixed axis through the angle $d\theta$ in a time interval dt is

$$\frac{dW}{dt} = \tau\frac{d\theta}{dt}.$$

Because dW/dt is the instantaneous power P delivered by the force and $d\theta/dt = \omega$, the expression reduces to

$$P = \frac{dW}{dt} = \tau\omega. \tag{7.45}$$

This expression is analogous to $P = Fv$ for linear motion, while the expression $dW = \tau d\theta$ is analogous to $dW = F_x dx$.

Furthermore, we can express the net torque as

$$\sum \tau = I\alpha = I\frac{d\omega}{dt} = I\frac{d\omega}{d\theta}\frac{d\theta}{dt} = I\frac{d\omega}{d\theta}\omega,$$

where we have made use of the chain rule in the third equality. We can rewrite the above expression in the following form:

$$\sum \tau d\theta = I\omega d\omega. \tag{7.46}$$

Integrating the work as the angular velocity changes from ω_i to ω_f, based on Eq. (7.46), we have

$$\sum W = \int_{\omega_i}^{\omega_f} I\omega d\omega = \frac{1}{2}I\omega_f^2 - \frac{1}{2}I\omega_i^2. \tag{7.47}$$

This is the *work-energy theorem* for rotational motion, which states that the net work done by external forces in rotating a rigid body about a fixed axis equals the change in the body's rotational energy.

7.3.4 Theoretical Framework for Rotational Dynamics with both Translational and Rotational Motion

In this section, we shall develop a theoretical framework that allows us to solve rotational dynamics problems that concern both translational and rotational motion without getting into details on inertia tensors and Euler's angles. The formulation demonstrates that dynamical motion can be decomposed into a translational and a rotational part: with the former captured by linear motion with the mass of the rigid body concentrated at the center-of-mass (COM) of the body[3], while the latter involves rotational motion of the extended body about its COM. The purpose of this section is to deduce the validity of this perspective. Before we do that, let us first introduce the idea of the Center-of-Mass frame.

7.3.4.1 Center-of-Mass Frame

A Center-of-Mass (COM) frame is a reference frame with its origin located at the COM of the extended body. It is in general a non-inertial reference frame because the body can execute accelerating motion.

---

[3]This is the particle representation we have been adopting in the analysis of translational motion.

Let us begin by uncovering certain properties of the COM frame through its description on the motion of a rigid body. We first reference the rigid body through an arbitrary point O which serves as the origin of an inertial reference frame. Based on this frame, the position vector of particle i of mass m_i relative to O is $\vec{r}_i$. Furthermore, $\vec{r}_{CM}$ is the position vector of the COM of this body with respect to O. It obeys the following relation based on Eq. (6.15).

$$M\vec{r}_{CM} = \sum_i m_i \vec{r}_i \,, \tag{7.48}$$

where $M = \sum_i m_i$. Next, we define a COM frame with origin O' that is attached to the COM of the body. Notably, we assume that all the axes of the COM frame are parallel to the reference frame of point O at all times[4]. With respect to the COM frame, the position vector of mass m_i is $\vec{r}_i'$. We expect these position vectors to satisfy the following vector equation (see Fig. 7.7):

$$\vec{r}_i = \vec{r}_{CM} + \vec{r}_i' \,. \tag{7.49}$$

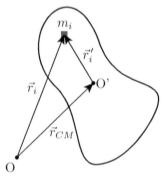

Fig. 7.7

Putting Eq. (7.49) into Eq. (7.48), we immediately observe that

$$M\vec{r}_{CM} = \sum_i m_i \left(\vec{r}_{CM} + \vec{r}_i'\right) = \left(\sum_i m_i\right)\vec{r}_{CM} + \sum_i m_i \vec{r}_i' \,,$$

---

[4]This will prevent the need to consider all the fictitious forces that are associated with the rotation of the COM frame axes relative to the inertial reference frame axes.

which implies that

$$\sum_i m_i \vec{r}_i' = 0 . \tag{7.50}$$

This equation indicates that the mass-weighted sum of the position vectors of all the individual particles of the extended body about its COM is zero. We next differentiate Eq. (7.50) with respect to time to obtain:

$$\frac{d\left(\sum_i m_i \vec{r}_i'\right)}{dt} = \sum_i m_i \frac{d\vec{r}_i'}{dt} = 0 ,$$

which shows that

$$\sum_i m_i \vec{v}_i' = 0 . \tag{7.51}$$

Note that $\vec{v}_i'$ is the velocity of particle i in the COM frame. Since $\vec{p}_i' = m_i \vec{v}_i'$, we have

$$\sum_i \vec{p}_i' = 0 . \tag{7.52}$$

Thus, the total linear momentum of all the particles of the body about the COM is zero.

By differentiating Eq. (7.51) with respect to time, we obtain

$$\sum_i m_i \vec{a}_i' = 0 , \tag{7.53}$$

where $\vec{a}_i'$ is the acceleration of particle i in the COM frame.

Let us now turn our attention to rotational motion. Since the COM frame is a non-inertial reference frame, our aim in the following evaluation is to determine the relation between the angular momentum $\vec{L}_{CM}$ and the torque $\vec{\tau}_{CM}$ with respect to the COM frame. We shall first perform our evaluation with respect to an arbitrary non-inertial reference frame by taking an alternative point A on the body as the origin. We then denote the position vector of A with respect to a particular inertial reference frame as $\vec{r}_A$. Similarly, we denote the position vector of particle i with respect to this same inertial reference frame as $\vec{r}_i$. Then, the angular momentum $\vec{L}_A$ of the rigid body with respect to the non-inertial reference frame A is given by

$$\vec{L}_A = \sum_i \vec{r}_{i/A} \times \vec{p}_{i/A} ,$$

$$= \sum_i (\vec{r}_i - \vec{r}_A) \times m_i (\vec{v}_i - \vec{v}_A) , \tag{7.54}$$

where the position vector $\vec{r}_{i/A}$ of particle i with respect to point A is given by $\vec{r}_i - \vec{r}_A$ and the associated linear momentum $\vec{p}_{i/A}$ of the same particle relative to A is given by $m_i(\vec{v}_i - \vec{v}_A)$. We now proceed to differentiate $\vec{L}_A$ with respect to time as follows:

$$\frac{d\vec{L}_A}{dt} = \frac{d}{dt}\left(\sum_i (\vec{r}_i - \vec{r}_A) \times m_i(\vec{v}_i - \vec{v}_A)\right) ,$$

$$= \sum_i (\vec{v}_i - \vec{v}_A) \times m_i(\vec{v}_i - \vec{v}_A) + \sum_i (\vec{r}_i - \vec{r}_A) \times m_i(\vec{a}_i - \vec{a}_A) ,$$

$$= \sum_i (\vec{r}_i - \vec{r}_A) \times m_i\vec{a}_i - \sum_i (\vec{r}_i - \vec{r}_A) \times m_i\vec{a}_A ,$$

$$= \sum_i (\vec{r}_i - \vec{r}_A) \times \vec{F}_{i,ext} - \sum_i m_i\vec{r}_i \times \vec{a}_A + \sum_i m_i\vec{r}_A \times \vec{a}_A ,$$

$$= \sum_i (\vec{r}_i - \vec{r}_A) \times \vec{F}_{i,ext} - M\vec{r}_{CM} \times \vec{a}_A + M\vec{r}_A \times \vec{a}_A ,$$

$$= \sum_i (\vec{r}_i - \vec{r}_A) \times \vec{F}_{i,ext} + (\vec{r}_{CM} - \vec{r}_A) \times (-M\vec{a}_A) , \tag{7.55}$$

where we have made use of: $(\vec{v}_i - \vec{v}_A) \times m_i(\vec{v}_i - \vec{v}_A) = 0$ according to the property of the vector product. Also, we have applied Newton's second law to yield $m_i\vec{a}_i = \vec{F}_{i,ext} + \vec{F}_{i,int}$, with $\vec{F}_{i,ext}$ being the net external force while $\vec{F}_{i,int}$ the net internal force on particle i. With the internal forces forming equal and opposite pairs, application of Newton's third law leads to pairs of internal torques canceling each other. Hence, only torques due to external forces remain. Finally, we have employed Eq. (7.48) and $\sum_i m_i = M$ in the above evaluation. Since $\vec{\tau}_A = \sum_i (\vec{r}_i - \vec{r}_A) \times \vec{F}_{i,ext}$, Eq. (7.55) can be rewritten as

$$\frac{d\vec{L}_A}{dt} = \vec{\tau}_A + (\vec{r}_{CM} - \vec{r}_A) \times (-M\vec{a}_A) . \tag{7.56}$$

Thus, Eq. (7.56) shows that in general, the expression $\vec{\tau}_A = d\vec{L}_A/dt$ does not apply when the reference frame is non-inertial. In particular, there exists an additional torque due to the fictitious force $-M\vec{a}_A$ which acts at the COM. However, if we were to locate A at the COM of the rigid body, the torque $(\vec{r}_{CM} - \vec{r}_A) \times (-M\vec{a}_A)$ vanishes. In consequence, the relation

$$\vec{\tau}_{CM} = \frac{d\vec{L}_{CM}}{dt} \tag{7.57}$$

is satisfied for the COM frame, even though the COM frame is a non-inertial reference frame.

In summary, Eqs. (7.50), (7.51), (7.52), (7.53) and (7.57) are physical relationships fulfilled by a COM frame.

7.3.4.2 Parallel Axis Theorem

In this section, we shall derive a relationship between the moment of inertia of a rigid body about an arbitrary reference point O and that relative to the COM. By definition, the moment of inertia about a point O (see Fig. 7.7) is given by

$$I_O = \sum_i m_i r_i^2 \,.$$

Noting that $r_i^2 = \vec{r}_i \cdot \vec{r}_i$, we have

$$I_O = \sum_i m_i \vec{r}_i \cdot \vec{r}_i \,. \tag{7.58}$$

Substituting Eq. (7.49) into Eq. (7.58)

$$I_O = \sum_i m_i \left(\vec{r}_{CM} + \vec{r}_i'\right) \cdot \left(\vec{r}_{CM} + \vec{r}_i'\right) \,,$$

$$= \sum_i m_i \left(\vec{r}_{CM} \cdot \vec{r}_{CM} + \vec{r}_{CM} \cdot \vec{r}_i' + \vec{r}_i' \cdot \vec{r}_{CM} + \vec{r}_i' \cdot \vec{r}_i'\right) \,,$$

$$= \sum_i m_i \vec{r}_{CM} \cdot \vec{r}_{CM} + 2\left(\sum_i m_i \vec{r}_i'\right) \cdot \vec{r}_{CM} + \sum_i m_i \vec{r}_i' \cdot \vec{r}_i' \,,$$

$$= M\vec{r}_{CM} \cdot \vec{r}_{CM} + \sum_i m_i \vec{r}_i'^2 \,,$$

where we have made use of $M = \sum_i m_i$ and Eq. (7.50). Since $\vec{r}_{CM} \cdot \vec{r}_{CM} = |\vec{r}_{CM}|^2 = R^2$ where R is the distance from the COM to the point O, and $I_{CM} = \sum_i m_i \vec{r}_i'^2$, we have

$$I_O = MR^2 + I_{CM} \,, \tag{7.59}$$

relating the moment of inertia of the object about point O to the moment of inertia of the same object about its COM. This equation is also known as the parallel axis theorem.

7.3.4.3 Development of the Theoretical Framework

We begin by developing the following set of equations. First, we differentiate both sides of Eq. (7.49) with resepect to time to obtain

$$\vec{v}_i = \vec{v}_{CM} + \vec{v}_i'. \tag{7.60}$$

Next, we differentiate Eq. (7.60) with respect to time and arrive at

$$\vec{a}_i = \vec{a}_{CM} + \vec{a}_i'. \tag{7.61}$$

Note that these equations serve to relate the velocity and acceleration of a particle measured with respect to the inertial reference frame to that measured relative to the COM frame. We next multiply both sides of Eq. (7.60) by m_i and using the definition of linear momentum, we obtain the following relation:

$$\vec{p}_i = m_i \vec{v}_{CM} + \vec{p}_i'. \tag{7.62}$$

We are now ready to develop the theoretical framework. By considering the angular momentum of the rigid object about O and making use of the equations established above, we perform the following series of mathematical evaluations:

$$\vec{L}_O = \sum_i \vec{r}_i \times \vec{p}_i,$$

$$= \sum_i (\vec{r}_{CM} + \vec{r}_i') \times (m_i \vec{v}_{CM} + \vec{p}_i'),$$

$$= \sum_i \vec{r}_{CM} \times m_i \vec{v}_{CM} + \sum_i \vec{r}_{CM} \times \vec{p}_i' + \sum_i \vec{r}_i' \times m_i \vec{v}_{CM} + \sum_i \vec{r}_i' \times \vec{p}_i',$$

$$= \vec{r}_{CM} \times \left(\sum_i m_i \right) \vec{v}_{CM} + \vec{r}_{CM} \times \left(\sum_i \vec{p}_i' \right) + \left(\sum_i m_i \vec{r}_i' \right) \times \vec{v}_{CM}$$

$$+ \sum_i \vec{r}_i' \times \vec{p}_i'.$$

Since the two terms within the bracket in the last line of the above equation are zero (see Eqs. (7.50) and (7.52)), and $\sum_i m_i = M$ and $\vec{L}_{CM} = \sum_i \vec{r}_i' \times \vec{p}_i'$, we have

$$\vec{L}_O = \vec{r}_{CM} \times M \vec{v}_{CM} + \vec{L}_{CM}. \tag{7.63}$$

Noting that

$$\vec{\tau}_O = \frac{d\vec{L}_O}{dt}, \tag{7.64}$$

and by determining $d\vec{L}_O/dt$ through differentiating Eq. (7.63) with respect to time, we have

$$\frac{d\vec{L}_O}{dt} = \frac{d\left(\vec{r}_{CM} \times M\vec{v}_{CM}\right)}{dt} + \frac{d\vec{L}_{CM}}{dt},$$

$$= \frac{d\vec{r}_{CM}}{dt} \times M\vec{v}_{CM} + \vec{r}_{CM} \times M\frac{d\vec{v}_{CM}}{dt} + \frac{d\vec{L}_{CM}}{dt}.$$

Now, the first term of the last line of the above equation vanishes because $d\vec{r}_{CM}/dt = \vec{v}_{CM}$ and $\vec{v}_{CM} \times M\vec{v}_{CM} = 0$. Furthermore, with $d\vec{v}_{CM}/dt = \vec{a}_{CM}$ and $d\vec{L}_{CM}/dt = \vec{\tau}_{CM}$ (see Eq. (7.57)), we obtain

$$\vec{\tau}_O = \vec{r}_{CM} \times M\vec{a}_{CM} + \vec{\tau}_{CM}. \tag{7.65}$$

Let us next multiply both sides of Eq. (7.61) by m_i and sum over all the particles:

$$\sum_i m_i \vec{a}_i = \sum_i m_i \vec{a}_{CM} + \sum_i m_i \vec{a}_i'.$$

This leads to

$$\sum_i m_i \vec{a}_i = M\vec{a}_{CM},$$

since $\sum_i m_i \vec{a}_i' = 0$ (see Eq. (7.53)). Here, we observe that

$$M\vec{a}_{CM} = \sum_i m_i \vec{a}_i = \vec{F}_{ext}, \tag{7.66}$$

where we have applied Newton's second law with respect to an inertial reference frame at the second equality. Equation (7.66) shows that even though the external force $\vec{F}_{ext}$ may not run through the COM, if we want to compute the acceleration of the COM ($\vec{a}_{CM}$), we can just assume it acts through the COM. Putting this result into Eq. (7.65), we yield

$$\vec{\tau}_O = \vec{r}_{CM} \times \vec{F}_{ext} + \vec{\tau}_{CM}. \tag{7.67}$$

With the reference point O being arbitrary, Eq. (7.67) implies a framework for the solution of rotational dynamics problems that involves both translational and rotational motion. Let us illustrate the application of this framework through a specific case. In Fig. 7.8(a), an extended rigid object is being acted upon by an external force $\vec{F}_{ext}$ at point A. Equation (7.67) indicates that the dynamics of this problem is equivalent to a shift of the point of action of the force $\vec{F}_{ext}$ from point A to the COM located at O'. In addition, this shift can only be considered legitimate unless we include the torque $\vec{\tau}_{CM}$, due to $\vec{F}_{ext}$ with respect to COM, at the same time. Note that

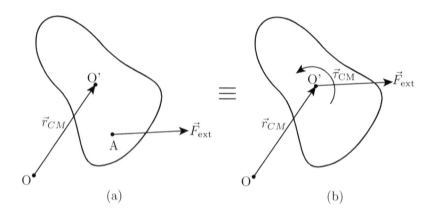

Fig. 7.8

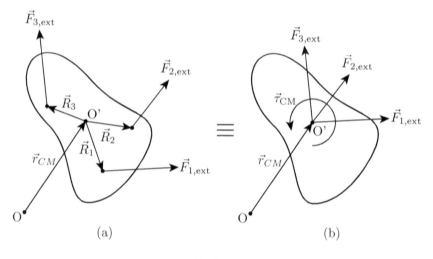

Fig. 7.9

$\vec{\tau}_{CM} = \vec{r}'_F \times \vec{F}_{ext}$, where $\vec{r}'_F$ is the position vector from the COM to the point of application of $\vec{F}_{ext}$, which is point A. In other words, the rotational dynamics that result from the action of $\vec{F}_{ext}$ as depicted in Fig. 7.8(a) is equivalent to that due to the action of the external force and torque given in Fig. 7.8(b).

In general, there could be more than one external force that act on the rigid body (see Fig. 7.9(a)). Applying our framework to this case, we have to shift each of the three external forces $\vec{F}_{1,ext}$, $\vec{F}_{2,ext}$, and $\vec{F}_{3,ext}$ to the COM at O'. At the same time, we have to add the torque contributed by each of these three external forces, i.e., $\vec{R}_1 \times \vec{F}_{1,ext}$, $\vec{R}_2 \times \vec{F}_{2,ext}$, and $\vec{R}_3 \times \vec{F}_{3,ext}$, to the system to compensate for the shift of each of these external forces (see Fig. 7.9(b)). More precisely, we have

$$\sum \vec{F}_{ext} = \vec{F}_{1,ext} + \vec{F}_{2,ext} + \vec{F}_{3,ext} = M\vec{a}_{CM} \qquad (7.68)$$

and

$$\sum \vec{\tau}_{CM} = \vec{R}_1 \times \vec{F}_{1,ext} + \vec{R}_2 \times \vec{F}_{2,ext} + \vec{R}_3 \times \vec{F}_{3,ext}, \qquad (7.69)$$

with $\vec{R}_i$ being the position vector from the COM to the point of application of $\vec{F}_{i,ext}$. With this, Eq. (7.67) takes the more general form:

$$\sum \vec{\tau}_O = \vec{r}_{CM} \times \sum \vec{F}_{ext} + \sum \vec{\tau}_{CM}. \qquad (7.70)$$

Based on the above discussion, we observe that the framework justifies our earlier use of the particle model, where all the external forces act on a particle, and here, all these forces act instead through the COM. From our framework, we observe that the particle in fact corresponds to the COM. In both pictures, all the mass of the body can be viewed as concentrated at the particle or the COM. Moreover, because the body is now extended, we have to include the torque on the system, and these torque are due to each of these forces relative to the COM, as if the COM is the fixed axis of rotation. In this way, our theoretical framework has separated the mechanics of rotational dynamics into a translational and a rotational part. The translational part with the resultant external force acting at the COM allows a set of equations due to Newton's second law on linear motion to be written. On the other hand, the external torque about the COM leads to another set of equations from Newton's second law applying to rotational motion. The solution of these two sets of equations together with a set of equations of constraint[5] of the system then depicts the full rotational dynamics of the system-of-interest. In the next section, we shall apply this theoretical framework to address the rotational dynamics of rolling motion.

Finally, let us show that the energy of the rigid body can also be split into a translational and a rotational part. The total energy of the body is

---

[5]The equation of constraint will be specifically illustrated through examples given in the later part of this chapter.

purely kinetic, and recalling Eq. (7.42), we have

$$K_R = \sum_i \frac{1}{2} m_i v_i^2 ,$$

$$= \frac{1}{2} I_O \omega^2 ,$$

$$= \frac{1}{2} \left(MR^2 + I_{CM} \right) \omega^2 ,$$

$$= \frac{1}{2} M \left(R\omega \right)^2 + \frac{1}{2} I_{CM} \omega^2 ,$$

$$= \frac{1}{2} M v_{CM}^2 + \frac{1}{2} I_{CM} \omega^2 , \tag{7.71}$$

where we have made use of the parallel-axis theorem (see Eq. (7.59)) at the third line of the equation. Furthermore, we have employed $|\vec{v}_{CM}| = v_{CM} = R\omega$, with R being the distance from point O to the COM of the body. Note that the total kinetic energy of the rigid body can be expressed as a sum of translational kinetic energy of the COM and rotational kinetic energy about the COM.

7.3.5 *Rolling Motion*

Rolling motion involves both translation and rotation, and the analysis of its dynamics provides a good illustration on the applicability of our framework.

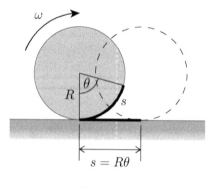

Fig. 7.10

Let us begin by defining the concept of *pure rolling motion*. Consider a uniform cylinder of radius R rolls on a horizontal surface. If this cylinder were to execute pure rolling motion, *it rolls without slipping*. This is a measurable effect. Under pure rolling motion, the cylinder rotates through an angle θ as its COM moves through a distance of $s = R\theta$ (see Fig. 7.10). Thus, the linear velocity of the cylinder is given by

$$v_{CM} = \frac{ds}{dt} = R\frac{d\theta}{dt} = R\omega \, . \tag{7.72}$$

The acceleration of the COM is

$$a_{CM} = \frac{dv_{CM}}{dt} = R\frac{d\omega}{dt} = R\alpha \, . \tag{7.73}$$

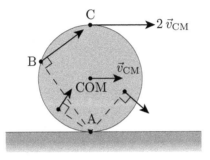

Fig. 7.11

As the cylinder rolls, there is a transition of the points on and within the cylinder to other parts of the body as illustrated in Fig. 7.11. Specifically, the point A on the cylinder will eventually rotate to point B and C. Note that the linear velocity of any point in or on the cylinder is in a direction perpendicular to the line from that point to the contact point A. Since no slipping occurs, the part of the rim that is at point A is at rest relative to the surface at the moment of contact. In other words, $v_{A/surface} = 0$.[6] In fact, rolling motion can be perceived as a combination of translational and rotational motion as shown in Fig. 7.12. In the figure, the part on translational motion has all the particles in the cylinder move with the same velocity v_{CM}. On the other hand, all the particles rotate about the COM with angular velocity ω as if the COM is a fixed axis for the rotational part. This combination of translational part and rotational part then gives rise to the rolling motion of the cylinder. In fact, by adding velocities at selected points on the cylinder, one immediately observes the consistency of the results. At the COM, we expect a velocity of v_{CM}. At point A, the corresponding velocities of the translational and rotational part cancel,

---

[6]Note that the surface may move. But as long as A moves along such that $v_{A/surface} = 0$, the condition holds.

leading to the expected zero velocity at this point. At the top rim of the cylinder, the velocities of the translational and rotational part add exactly, leading to $v = v_{CM} + R\omega = 2v_{CM}$, where we have used $v_{CM} = R\omega$ because of pure rolling motion. The rest of the velocities of the rolling cylinder can also be similarly determined, by using vector addition of the velocities of the translational and rotational part at respective points.

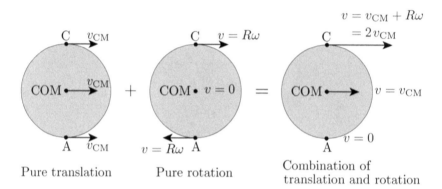

| Pure translation | Pure rotation | Combination of translation and rotation |

Fig. 7.12

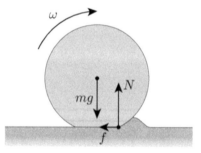

Fig. 7.13

For a pure rolling cylinder that is already translating with constant velocity v_{CM} on a horizontal surface, there is no need for a driving force to sustain its motion for a perfectly rigid body. The reason can be understood by applying our framework as follows. The external forces that act on such a cylinder is basically the gravitational force exerted on it from the Earth, and also the normal force from the surface of the ground on which it rolls. Since these forces cancel at the COM and they do not produce any torque about the COM, there is no a_{CM} and α to change the rolling motion. Hence, the cylinder should maintain its motion regardless of whether the surface is rough or frictionless. One may question whether friction should be acting in such a situation. Since the contact point of the cylinder on the surface has zero relative velocity due to pure rolling motion, the friction cannot be kinetic friction. If it is static friction, it could only act directly to the right or to the left. It cannot act to the right because it will cause the cylinder to accelerate in its translational motion, which is against our experience. If the static friction were to point to the left, it will cause the cylinder to increase its rotational speed in the clockwise direction, which is a phenomenon yet to be observed. Thus, the frictional force cannot be static friction as well. However, our experience tells us that the cylinder with pure rolling motion does decelerate and stop moving eventually on the ground. The cause is that of *rolling friction*. In the real-world, no object is perfectly rigid, and every object is elastic to some extent. The elastic nature of the cylinder results in its contact with the surface in the form of an extended area instead of a point. On this area, a force due to rolling friction directs to the left and upward as shown in Fig. 7.13. It leads to an a_{CM} that retards the translational motion of the cylinder and an anti-clockwise torque that slows the cylinder's rotational speed. This explains why a pure rolling cylinder stop rolling in the real-world.

7.3.6   *Examples on Rolling Motion*

7.3.6.1   *Rolling Down an Inclined Plane*

In this example, an object rolls without slipping down an inclined plane, which is inclined at an angle β to the horizontal (see Fig. 7.14(a)). We will determine the object's acceleration and the magnitude of the friction force acting on it. In addition, we will evaluate the pseudowork done by the static friction on the object and show that the static friction serves to convert part of the object's translational kinetic energy to the object's rotational kinetic energy.

A figure of the problem is illustrated in Fig. 7.14(a) where we show the point of application of the external forces on the object: static friction and normal force from the inclined plane, and gravitational force from Earth. By means of our theoretical framework of Section 7.3.4, we need to shift these external forces to the COM of the object while at the same time, include the torque of these forces about the COM. The outcome of these steps are shown in Fig. 7.14(b). The first step allows us to write the

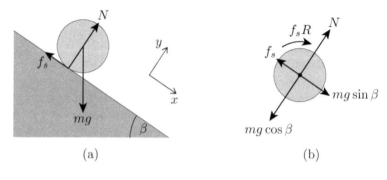

Fig. 7.14

following set of equations relating to the translational motion of the object using Newton's second law:

$$N = mg\cos\beta, \tag{7.74}$$

$$mg\sin\beta - f_s = m\,a_{CM}. \tag{7.75}$$

Applying Newton's second law on rotation to the second step, we have

$$f_s R = I_{CM}\alpha. \tag{7.76}$$

Note that only the static friction force contributes to the torque about the COM. As the object is executing pure rolling motion, its motion needs to satisfy the following condition:

$$a_{CM} = R\alpha. \tag{7.77}$$

Such a condition is known as the equation of constraint for the problem.

After stating all the necessary equations of the problem, we are now ready to determine its solution. Putting α of Eq. (7.77) into Eq. (7.76), we have

$$f_s R = I_c m \left(\frac{a_{CM}}{R}\right), \tag{7.78}$$

which gives

$$f_s = \frac{I_{CM}\,a_{CM}}{R^2}.$$

Substituting this expression of f_s into Eq. (7.75), we proceed to solve for a_{CM}:

$$mg\sin\beta - \frac{I_{CM}\,a_{CM}}{R^2} = m\,a_{CM},$$

which leads to

$$a_{CM} = \frac{mgR^2 \sin \beta}{mR^2 + I_{CM}} \, . \tag{7.79}$$

Inserting this result into Eq. (7.78), we obtain the static friction force:

$$f_s = \frac{mgI_{CM} \sin \beta}{mR^2 + I_{CM}} \, . \tag{7.80}$$

It is interesting that static friction is required for pure rolling motion along an inclined plane. Recall that static friction is not required for pure rolling motion of an object on horizontal ground as discussed in section 7.3.5. The purpose of static friction in the present case can be understood by determining its pseudowork done on the object as the object rolls down the inclined plane.

First, let us note that the object rolls a distance d down the length of the inclined plane, where $d = h/\sin \beta$ with h being the height of the inclined plane. Since the object starts from rest, its COM velocity v_{CM} at the bottom of the inclined plane obeys

$$\begin{aligned} v_{CM}^2 &= 2a_{CM}d^2 \, , \\ &= 2\left(\frac{mgR^2 \sin \beta}{mR^2 + I_{CM}}\right)\left(\frac{h}{\sin \beta}\right) \, , \\ &= \frac{2mghR^2}{mR^2 + I_{CM}} \, . \end{aligned} \tag{7.81}$$

Thus, the translational kinetic energy of the object at the bottom of the inclined plane is

$$K_T = \frac{1}{2}mv_{CM}^2 = \frac{m^2ghR^2}{mR^2 + I_{CM}} \, . \tag{7.82}$$

Analogously, we can determine the rotational kinetic energy of the object at the bottom of the inclined plane K_R. Noting that the angular acceleration of the object

$$\alpha = \frac{a_{CM}}{R} = \frac{mgR \sin \beta}{mR^2 + I_{CM}} \, ,$$

and its angular displacement down the inclined plane is

$$\phi = \frac{h}{R \sin \beta} \, ,$$

we yield the following relation for its angular velocity at the bottom of the inclined plane:

$$\omega^2 = 2\alpha\phi\,,$$
$$= 2\left(\frac{mgR\sin\beta}{mR^2 + I_{CM}}\right)\left(\frac{h}{R\sin\beta}\right)\,,$$
$$= \frac{2mgh}{mR^2 + I_{CM}}\,.$$

Thus,

$$K_R = \frac{1}{2}I_{CM}\omega^2 = \frac{mghI_{CM}}{mR^2 + I_{CM}}\,. \tag{7.83}$$

Summing Eqs. (7.82) and (7.83), we notice that

$$K_T + K_R = \frac{m^2ghR^2}{mR^2 + I_{CM}} + \frac{mghI_{CM}}{mR^2 + I_{CM}}\,,$$
$$= \frac{mgh\left(mR^2 + I_{CM}\right)}{mR^2 + I_{CM}}\,,$$
$$= mgh\,. \tag{7.84}$$

This result affirms the *Conservation of Energy* of the system as the object starts with a potential energy of mgh before rolling. The static friction that acts on the object does not dissipate the energy because it performs zero real work. Let us now determine the pseudowork done by f_s along the inclined plane as it acts on the object:

$$W = f_s d\,,$$
$$= \left(\frac{mgI_{CM}\sin\beta}{mR^2 + I_{CM}}\right)\left(\frac{h}{\sin\beta}\right)\,,$$
$$= \frac{mghI_{CM}}{mR^2 + I_{CM}}\,,$$
$$= K_R\,. \tag{7.85}$$

Noting that the torque of the static friction about the COM is $\tau_{CM} = f_s R$, $W = f_s d = f_s R\phi = \tau_{CM}\phi$. Thus, the static friction acts by transferring $W = K_R$ amount of rotational energy to the object by means of rotational pseudowork. Where does this energy comes from? It comes from the translational part because f_s acts directly opposite to the direction of motion of the object, leading to a negative pseudowork of $-W = -f_s d$ done on the translational mode of motion. Notice that this negative pseudowork done has exactly the same magnitude as the pseudowork done that has been transferred to the rotational mode of motion. Without static friction, such a transfer of energy from the translational mode to the rotational mode of motion will not be possible as exemplified when the inclined plane is frictionless.

7.3.6.2 Rolling Up an Inclined Plane

Assume now that the same object of Section 7.3.6.1 moves up the inclined plane of angle β with initial center of mass velocity v_{CM} and angular velocity ω. Let us analyze this situation based on energy consideration. If the inclined plane is frictionless, the object would translate up the slope as it slides and rotates with a constant angular velocity ω. In this case, the only force that does work is $-mg\sin\beta$, which acts down the inclined plane. When the object reaches the highest point of the slope, the work done by this force has converted all the translational kinetic energy of the object to gravitational potential energy after traveling a distance of $d_1 = v_{CM}^2/(2g\sin\beta)$. At this highest point when the object stops translating, it is nonetheless still rotating with ω, because there is no friction acting on the object to generate a torque to change its angular velocity, and hence its rotational kinetic energy $I\omega^2/2$ remains constant throughout its motion up the plane.

Now let's suppose friction is present. It is interesting that the pseudowork of static friction can tap into this rotational kinetic energy of the object by mediating its transfer to the form of translational kinetic energy, whereupon it is further converted to gravitational potential energy. This effect can be observed from the dynamics of the object when there is friction on the inclined plane. Here, the object moves with pure rolling motion up the slope with the additional force of static friction acting in the direction of motion at the point where the object comes in contact with the plane. The static friction performs zero real work because its point of application does not move. Instead, it performs pseudowork on the object by converting its rotational kinetic energy to translational kinetic energy. Specifically, the static friction exerts a counter-torque of $f_s R$ against the rotation of the object, which draws an energy of $f_s R\Delta\theta_2$ from the rotational mode. Note that $\Delta\theta_2$ is the total angular displacement of the object from the beginning of its motion to the end when it reaches the highest point of the slope. This energy is transferred to the translational mode via a positive pseudowork of $f_s d_2$, where $d_2 = R\Delta\theta_2$ is the corresponding total linear distance traveled by the object. Because $f_s R = I\alpha$ and $a_{CM} = R\alpha$, we have $f_s = Ia_{CM}/R^2$. Using the kinematic equation with constant a_{CM} up the slope, we obtain $d_2 = v_{CM}^2/(2a_{CM})$. This leads to $f_s d_2 = I v_{CM}^2/(2R^2)$, which is the additional translational kinetic energy of the object derived through the pseudowork of f_s. Thus, pseudowork plays the important role of mediating the inter-conversion of kinetic energy within the object, with the consequence that the object gains a greater gravitational potential energy as it travels a (counter-intuitively) *farther* distance of $\Delta d = f_s d_2/(mg \sin \beta) = I v_{CM}^2/(2mgR^2 \sin \beta)$ up the slope relative to the frictionless case. In other words, we have $d_2 = d_1 + \Delta d$, or $d_2 > d_1$.

7.3.6.3 *Pure Rolling Motion of an Accelerating Car Wheel*

Here, we consider a car initially at rest on the road. It then starts to accelerate with constant acceleration. We assume that the mass and radius of a wheel of the car is M and R respectively, with the moment of inertia of the wheel about its COM being I_{CM}. We will first determine the acceleration of the wheel if it is driven by a torque τ_e from the engine of the car. Note that the wheel undergoes pure rolling motion as the car moves. After which, we will show that the pseudowork done by static friction F_s from the road serves to transfer part of the rotational kinetic energy of the rolling wheel to its translational kinetic energy.

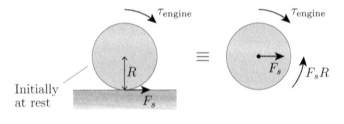

Fig. 7.15

Again, we first establish a set of translational and rotational equation of motions by applying Newton's second law to the equivalent model (refer to the right-hand side of Fig. 7.15) developed from our theoretical framework:

$$F_s = M a_{CM}, \tag{7.86}$$

$$\tau_e - F_s R = I_{CM} \alpha, \tag{7.87}$$

$$a_{CM} = R\alpha. \tag{7.88}$$

Substituting $\alpha = a_{CM}/R$ into Eq. (7.87) and also employing Eq. (7.86), we have

$$\tau_e - Ma_{CM}R = \frac{I_{CM}\, a_{CM}}{R},$$

$$a_{CM}\left(MR + \frac{I_{CM}}{R}\right) = \tau_e.$$

Therefore,

$$a_{CM} = \frac{R\tau_e}{MR^2 + I_{CM}}. \tag{7.89}$$

Assuming that τ_e is constant, let us find the velocity of the COM v_{CM} of the wheel and also its angular velocity ω after the wheel has traveled a horizontal distance d. Since the wheel starts from rest, we have

$$v_{CM}^2 = 2a_{CM}d = 2\left(\frac{R\tau_e}{MR^2 + I_{CM}}\right)d.$$

Therefore, the wheel's translational kinetic energy after traveling the distance d is

$$K_T = \frac{1}{2}Mv_{CM}^2 = \frac{MRd\tau_e}{MR^2 + I_{CM}}. \tag{7.90}$$

On the other hand, let us now determine the rotational quantity $\Delta\theta$, which is the angular displacement of the wheel as it travels a linear distance of d as follows:

$$\Delta\theta = \frac{d}{R}.$$

Its angular velocity is then given by

$$\omega^2 = 2\alpha\Delta\theta = 2\left(\frac{a_{CM}}{R}\right)\Delta\theta = 2\left(\frac{\tau_e}{MR^2 + I_{CM}}\right)\left(\frac{d}{R}\right).$$

Consequently, the wheels' rotational kinetic energy at distance d from its starting point is

$$K_R = \frac{1}{2}I_{CM}\omega^2 = \frac{I_{CM}d\tau_e}{MR^3 + I_{CM}R}. \tag{7.91}$$

Summing Eqs. (7.90) and (7.91), we obtain

$$K_T + K_R = \frac{MRd\tau_e}{MR^2 + I_{CM}} + \frac{I_{CM}d\tau_e}{MR^3 + I_{CM}R},$$

$$= \frac{MR^2d\tau_e + I_{CM}d\tau_e}{MR^3 + I_{CM}R},$$

$$= \frac{d\tau_e \left(MR^2 + I_{CM}\right)}{R\left(MR^2 + I_{CM}\right)},$$

$$= \tau_e \frac{d}{R},$$

$$= \tau_e \Delta\theta. \tag{7.92}$$

This result shows that the rotational work done by the car engine has been transferred to the translational and rotational kinetic energy of the wheel. In particular, we note that the pseudowork done by the static friction on the wheel along the distance d:

$$W = F_s d,$$

$$= \frac{MRd\tau_e}{MR^2 + I_{CM}},$$

$$= \frac{1}{2}Mv_{CM}^2,$$

$$= K_T. \tag{7.93}$$

In addition, notice that the counter-clockwise torque due to F_s performs a negative pseudowork against that perform by the torque τ_e of the car engine. This negative pseudowork done $-F_s R\Delta\theta = -F_s d = -W$ has the same magnitude as that transferred to the translational motion of the wheel. Here, we observe that the static friction acts to mediate the transfer of an amount of rotational kinetic energy from the rotational work done by the car engine to the translational kinetic energy of the wheel. Imagine that the surface of the road is frictionless, such a transfer from rotational mode to translational mode of energy will not be possible. This is observed when a car is trapped in a slippery ditch, with the kinetic energy of the wheel being entirely in the rotational form, since there is no static friction to mediate a transfer to the translational form.

7.3.6.4   *Placing a Rotating Sphere onto a Ground with Friction*

A uniform solid sphere of mass m and radius R is set into motion with an angular speed ω_0. At time $t = 0$, the solid sphere is very gently dropped onto the horizontal surface, so that the vertical center of mass velocity just prior to touching the surface is almost zero. There is friction between the sphere and the surface. We will be working out three aspects of this example. First, we will determine the angular speed of rotation when the sphere finally rolls without slipping at time $t = T$. Second, we will evaluate the amount of kinetic energy lost by the sphere between $t = 0$ and $t = T$.

Finally, we will show that the result in the second part equals the work done against the frictional force that acts to cause the sphere to roll without slipping.

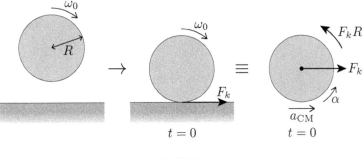

Fig. 7.16

The scenario of this example is drawn in Fig. 7.16 together with its formulation based on our theoretical framework. Writing down Newton's second law for the translational and rotational motion as the sphere touches the ground, we have

$$F_k = ma_{CM} \,, \tag{7.94}$$

$$F_k R = I_{CM}\alpha \,, \tag{7.95}$$

where F_k is the kinetic friction from the ground since there is relative motion between sphere and ground as a consequence of the sphere's rotational motion. In addition, the initial $v_{CM,i}$ of the sphere is zero just at the instance it touches the ground and therefore $v_{CM,i} < R\omega_0$ and there is no pure rolling motion at $t = 0$. In fact, there is rolling with slipping. Now, let the COM velocity of the sphere and its angular velocity at $t = T$ be v_{CM} and ω respectively; we have

$$v_{CM} = v_{CM,i} + a_{CM}T \,,$$

or

$$v_{CM} = a_{CM}T \,. \tag{7.96}$$

Furthermore, we have

$$\omega = \omega_0 - \alpha T \,. \tag{7.97}$$

Note that the formulation of Eqs. (7.96) and (7.97) results from both a_{CM} and α being constant. At $t = T$, the sphere rolls with pure rolling motion, which implies that it satisfies the condition:

$$v_{CM} = R\omega \,.$$

Substituting Eqs. (7.96) and (7.97) into this condition, we deduce that

$$a_{CM}T = R\omega_0 - R\alpha T \,. \tag{7.98}$$

From Eq. (7.94) and Eq. (7.95), we obtain $ma_{CM}R = I_{CM}\alpha$, which gives

$$a_{CM} = \frac{I_{CM}\alpha}{mR} \,. \tag{7.99}$$

Putting this into Eq. (7.98), we arrive at

$$\frac{I_{CM}}{mR}\alpha T = R\omega_0 - R\alpha T \,,$$

which leads to

$$\alpha T = \frac{R\omega_0}{R + \frac{I_{CM}}{mR}} \,. \tag{7.100}$$

By substituting this result into Eq. (7.97) and noting that $I_{CM} = 2mR^2/5$ for a solid sphere, we obtain

$$\omega = \omega_0 - \frac{R\omega_0}{R + \frac{I_{CM}}{mR}} \,,$$

$$= \omega_0 - \frac{R\omega_0}{R + \frac{2}{5}R} \,,$$

$$= \omega_0 - \frac{5}{7}\omega_0 \,,$$

$$= \frac{2}{7}\omega_0 \,. \tag{7.101}$$

This completes the first part.

For the second part, we will need to work out the initial kinetic energy K_i of the sphere and its final kinetic energy K_f at $t = T$ as follows:

$$K_i = \frac{1}{2}I_{CM}\omega_0^2 \,,$$

$$= \frac{1}{2}\left(\frac{2}{5}mR^2\right)\omega_0^2 \,,$$

$$= \frac{1}{5}mR^2\omega_0^2 \,,$$

and

$$K_f = \frac{1}{2}I_{CM}\omega^2 + \frac{1}{2}mv_{CM}^2 \,,$$

$$= \frac{1}{2}\left(\frac{2}{5}mR^2\right)\omega^2 + \frac{1}{2}mR^2\omega^2 \,,$$

$$= \frac{7}{10}mR^2\omega^2 \,,$$

$$= \frac{7}{10}mR^2\left(\frac{2}{7}\omega_0\right)^2 \,,$$

$$= \frac{4}{70}mR^2\omega_0 \,.$$

The kinetic energy lost is given by $K_i - K_f = (1/5 - 4/70)\,mR^2\omega_0 = mR^2\omega_0/7$.

In the final part of this example, we need to consider two scenarios on the displacement of the sphere due to the presence of slipping. The first scenario relates to linear displacement:

$$s = \frac{1}{2}a_{CM}T^2 = \frac{1}{2}\left(\frac{I_{CM}\alpha}{mR}\right)T^2 \,, \tag{7.102}$$

while the second scenario involves angular displacement:

$$\Delta\theta = \omega_0 T - \frac{1}{2}\alpha T^2 \tag{7.103}$$

which corresponds to the following linear displacement:

$$d = R\Delta\theta = R\omega_0 T - \frac{1}{2}R\alpha T^2 \,. \tag{7.104}$$

Because of slipping, $s \neq d$. In fact, the linear displacement that slips between the sphere and the ground is $(d - s)$, and we expect the work done by friction along this displacement to be the source of heat and frictional loss. Nonetheless, interesting physical insights can be gained by understanding each component of the work done by friction. For this, we shall perform the following sequence of evaluations to decipher its underlying meaning. Noting that

$$F_k = ma_{CM} = m\left(\frac{I_{CM}\alpha}{mR}\right) = \frac{I_{CM}\alpha}{R} \,,$$

we obtain

$$F_k s = \left(\frac{I_{CM}\alpha}{R}\right)\left(\frac{I_{CM}\alpha T^2}{2mR}\right) = \frac{I_{CM}^2\alpha^2 T^2}{2mR^2} \,. \tag{7.105}$$

Equation (7.105) gives the pseudowork done by F_k since the point of application of F_k did not move during the displacement s. Noting from Eq. (7.96) that $v_{CM} = a_{CM}T = I_{CM}\alpha T/mR$ and

$$v_{CM}^2 = \frac{I_{CM}^2 \alpha^2 T^2}{m^2 R^2} \,,$$

we observe that

$$F_k s = \frac{1}{2} m v_{CM}^2 \,, \tag{7.106}$$

which shows that this component of the work done by kinetic friction, i.e. $F_k s$, gives rise to the translational kinetic energy of the sphere. This is to be expected of pseudowork. Let us next evaluate $F_k d$ as follows:

$$F_k d = \frac{I_{CM}\alpha}{R} \left(R\omega_0 T - \frac{1}{2} R\alpha T^2 \right) \,,$$

$$= I_{CM} \left(\omega_0 \alpha T - \frac{1}{2} \alpha^2 T^2 \right) \,. \tag{7.107}$$

Before we can determine what Eq. (7.107) represents, we first square both sides of Eq. (7.96) as follows:

$$\omega^2 = \omega_0^2 - 2\omega_0 \alpha T + \alpha^2 T^2 \,,$$

$$\omega_0^2 - \omega^2 = 2\omega_0 \alpha T - \alpha^2 T^2 \,,$$

$$\frac{1}{2}\omega_0^2 - \frac{1}{2}\omega^2 = \omega_0 \alpha T - \frac{1}{2}\alpha^2 T^2 \,. \tag{7.108}$$

Substituting Eq. (7.108) into Eq. (7.107), we obtain

$$F_k d = \frac{1}{2} I_{CM}\omega_0^2 - \frac{1}{2} I_{CM}\omega^2 \,. \tag{7.109}$$

Equation (7.109) clearly shows that $F_k d$ takes away part of the initial rotational energy such that the remaining rotational energy on the sphere is $I_{CM}\omega^2/2$. A portion of the energy taken by $F_k d$ is converted to translational kinetic energy of the sphere via the pseudowork $F_k s$, with the remaining being dissipated as heat via the performance of real work. This fact is easily verified as follows:

$$F_k d - F_k s = \frac{1}{2} I_{CM}\omega_0^2 - \frac{1}{2} I_{CM}\omega^2 - \frac{1}{2} m v_{CM}^2 \,,$$

$$= \frac{1}{2} I_{CM}\omega_0^2 - \left(\frac{1}{2} I_{CM}\omega^2 + \frac{1}{2} m v_{CM}^2 \right) \,,$$

$$= K_i - K_f \,,$$

$$= \frac{1}{7} m R^2 \omega_0^2 \,,$$

which is the same as that obtained in part 2.

There is an alternative method to obtain the angular velocity of the sphere when it executes pure rolling motion at $t = T$. In this approach, we choose the reference O appropriately so that no external torque is present throughout the rotational dynamics of the sphere. The way to achieve this is to fix O on the surface of the ground as shown in Fig. 7.17. We shall now demonstrate that this is indeed the case. At $t = 0$ which is shown in Fig. 7.17(a), there is obviously no torque from $\vec{F}_k$ and the normal force $\vec{N}$ about O since $\vec{r}$ from O to the point of application of these forces is zero. For $m\vec{g}$, $\vec{r}$ points in a direction opposite to $m\vec{g}$, and hence $\vec{r} \times m\vec{g} = 0$. There is thus no external torque in this case. During the period $0 < t < T$ as the sphere is slipping as shown in Fig. 7.17(b), the vector $\vec{r}$ is parallel to $\vec{F}_k$, and therefore $\vec{r} \times \vec{F}_k = 0$. On the other hand, $\vec{r} \times \vec{N} = |\vec{r}||\vec{N}| \sin(\pi/2)\hat{k} = |\vec{r}||\vec{N}|\hat{k}$, where $\hat{k}$ is a unit vector pointing out of the page of the book. For $m\vec{g}$, we have $\vec{r'} \times m\vec{g} = -|\vec{r'}||m\vec{g}| \sin\phi\,\hat{k} = -|\vec{r'}||m\vec{g}| \sin(\pi/2 - \theta)\,\hat{k} = -|\vec{r'}||m\vec{g}| \sin\theta\,\hat{k} = -|\vec{r}||m\vec{g}|\hat{k}$ (see Fig. 7.17(b)) because $|\vec{r}| = |\vec{r'}| \sin\theta$. Since $|\vec{N}| = |m\vec{g}|$, we have $\vec{r} \times \vec{N} + \vec{r'} \times m\vec{g} = 0$. Therefore, there is also no external torque in these situations. A similar analysis for the period $t \geq T$ (see Fig. 7.17(c)) will also show an absence of external torque.

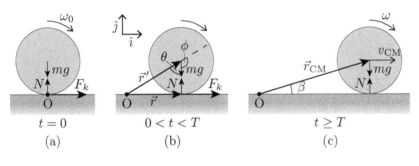

$$t = 0 \qquad\qquad 0 < t < T \qquad\qquad t \geq T$$
$$\text{(a)} \qquad\qquad\qquad \text{(b)} \qquad\qquad\qquad \text{(c)}$$

Fig. 7.17

Without external torque, we can apply the principle of *Conservation of Angular Momentum* to the problem. Using Eq. (7.63):

$$\vec{L}_O = \vec{r}_{CM} \times m\vec{v}_{CM} + \vec{L}_{CM} \,,$$

we have

$$\vec{L}_0^{initial} = \vec{L}_{CM}^{initial} = -I_{CM}\omega_0\hat{k} \qquad (7.110)$$

because v_{CM} at the initial instant is zero. Next, we have

$$\vec{L}_0^{final} = \vec{r}_{CM} \times m\vec{v}_{CM} + \vec{L}_{CM}^{final},$$
$$= -\left(mv_{CM}R + I_{CM}\omega\right)\hat{k}, \qquad (7.111)$$

where we have used the fact that $\vec{r}_{CM} \times m\vec{v}_{CM} = -|\vec{r}_{CM}||m\vec{v}_{CM}|\sin\beta\,\hat{k} = -|m\vec{v}_{CM}|R\hat{k}$ (see Fig. 7.17(c)). Applying conservation of angular momentum, we have

$$\vec{L}_0^{initial} = \vec{L}_0^{final},$$
$$I_{CM}\omega_0 = mv_{CM}R + I_{CM}\omega,$$
$$I_{CM}\omega_0 = mR^2\omega + I_{CM}\omega,$$

where we have employed the condition $v_{CM} = R\omega$ at the last line of the above equation due to pure rolling motion. Solving for ω, we obtain

$$\omega = \frac{I_{CM}\omega_0}{mR^2 + I_{CM}} = \frac{2}{7}\omega_0,$$

which is the same as Eq. (7.101).

7.3.6.5   Vehicles Turning in a Circle

We consider two types of vehicles: a rider on a motorcycle, and a car. First, we treat the system of rider-motorcycle executing a circular turn around a corner of the road. Based on experience, the motorcyclist would lean towards the corner as it turns. This is captured in the diagram illustrated in Fig. 7.18, where the rider-motorcycle is shown leaning at an angle θ away from the vertical. Because the motorcyclist is turning to the left, the rider-motorcycle system leans to the left, as depicted in the figure. The physical reason for this can be understood by performing an analysis using the theoretical framework we have earlier developed.

Fig. 7.19 shows the external forces acting on the rider-motorcycle system. There is the weight of the system $M\vec{g}$ acting vertically downward from its COM. The road exerts a normal force $\vec{N}$ and a static frictional force $\vec{F}_s$ at the contact point of the vehicle to the road. While $\vec{N}$ acts vertically upward, $\vec{F}_s$ is horizontal and points in the direction towards the center of the turn. As before, we simplify notation by writing these vectorial forces as Mg, N and F_s.

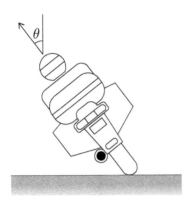

Fig. 7.18

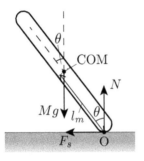

Fig. 7.19

Using our framework, we first shift N and F_s to the COM. To account
for this change, we compensate by introducing two torques about the COM.
The first torque τ_N acts counter-clockwise with a magnitude of $Nl_m \sin\theta$,
while the second torque τ_F acts clockwise with a value of $F_s l_m \cos\theta$. Note
that l_m is the distance from the contact point to the COM of the rider-
motorcycle system (see Fig. 7.19). Applying Newton's second law to the
equivalent model given in Fig. 7.20 with respect to an inertial reference
frame (such as the Earth), we have:

$$N = Mg, \tag{7.112}$$

$$F_s = MR\omega^2, \tag{7.113}$$

where R is the radius of the circle of the turn with center O_m and ω is
the angular velocity of the rider-motorcycle system about O_m. Eq. (7.113)
indicates that the centripetal force of the circular motion is attributed to

Based on framework

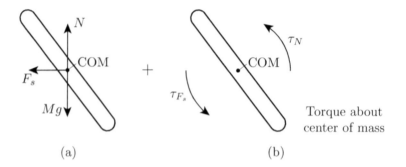

(a)   +   (b)

Torque about
center of mass

Fig. 7.20

the static friction. Assuming the rider-motorcycle system turns steadily
without toppling, τ_N and τ_F must counteract and cancel each other. Thus,

$$Nl_m \sin\theta - F_s l_m \cos\theta = 0 ,$$ (7.114)

which implies that

$$\tan\theta = \frac{F_s}{N} .$$ (7.115)

Equations (7.112) and (7.113) then give $\tan\theta = R\omega^2/g$. With $v = R\omega$, we
have

$$\theta = \tan^{-1}\left(\frac{v^2}{Rg}\right) .$$ (7.116)

This result shows that when the motorcycle turns with a larger speed v or
a smaller radius R of the circular path, the leaning angle θ becomes larger.
Furthermore, Eq. (7.115) indicates that the larger the F_s, the larger the
lean angle. However, there is a maximum static friction $F_{s,max} = \mu_s N$,
beyond which the motorcycle will slip. There is thus a maximum lean
angle θ_m which occurs at $F_{s,max}$. From Eq. (7.115), we have

$$\tan\theta_m = \frac{F_{s,max}}{N} ,$$
$$= \mu_s .$$

Now, if the rider has not intentionally leaned as it turns but instead
adopts the posture as shown in Fig. 7.21, the rider-motorcycle system
will topple and roll over. In this situation, there exists only the torque
$\tau_F = F_s l_m$ that turns the rider-motorcycle system clockwise, without the

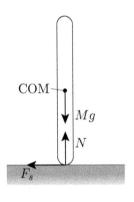

Fig. 7.21

presence of the counteracting torque τ_N. The net torque τ_F prevents the rider-motorcycle system from maintaining equilibrium as it executes the turn.

In the second case of consideration where a car turns around a corner, you would observe three clockwise torques due to F_{sA}, F_{sB}, and N_A, with their resultant torque counteracted by the anti-clockwise torque due to N_B. Unlike the motorcyclist which needs to lean to create the anti-clockwise torque from N to cancel the clockwise torque from F_s, the car already possesses the necessary elements for rotational equilibrium without leaning as it turns around the corner.

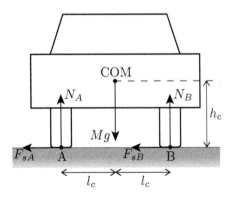

Fig. 7.22

The external forces that act on the car as it executes the circular path is shown in Fig. 7.22. In addition to Mg that exerts vertically downward at the COM of the car, there are forces that act on each wheel of the car. In our analysis, we treat the two inner wheels (the left wheels) as one wheel A, and the two outer wheels (the right wheels) as one wheel B. The forces N_A and F_{sA} are the normal and static frictional force that act on wheel A respectively, while N_B and F_{sB} are the corresponding forces that act on wheel B. Furthermore, h_c is the vertical height of the COM from the road and l_c is the horizontal distance from the COM to each wheel (see Fig. 7.22).

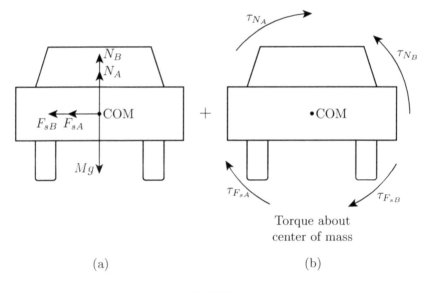

Torque about
center of mass

(a)   (b)

Fig. 7.23

As before, we apply our theoretical framework to study the dynamics of the car as it turns the corner. We first move N_A, F_{sA}, N_B and F_{sA} to the COM of the car. Next, we deduce the torques about the COM to ensure that the transformed model is dynamically equivalent to the original. The equivalent model is displayed in Fig. 7.23, where we indeed observe the presence of the three clockwise torques τ_{N_A}, $\tau_{F_{sA}}$, $\tau_{F_{sB}}$ and the anti-clockwise torque τ_{N_B} described earlier.

Applying Newton's second law on the equivalent model with respect to an inertial reference frame, we obtain

$$N_A + N_B = Mg\,, \tag{7.117}$$

$$F_{sA} + F_{sB} = MR\omega^2\,, \tag{7.118}$$

where the sum of the two static frictional forces contributes to the centripetal force for the turn, assuming that both wheels of the car maintain contact with the road. Here, R is the distance of the COM to the center O_c of the circular path and ω is the angular velocity of the car about O_c. Taking moment about COM and assuming that the car maintains rotational equilibrium, we expect

$$N_A l_c + F_{sA} h_c + F_{sB} h_c = N_B l_c . \tag{7.119}$$

In other words, the magnitude of the sum of the clockwise torques is equal to that of the anticlockwise torque. Substituting Eq. (7.118) into Eq. (7.119), we arrive at

$$N_B - N_A = \frac{MR\omega^2 h_c}{l_c} . \tag{7.120}$$

We then solve for N_A and N_B from Eqs. (7.117) and (7.120) to obtain

$$N_A = \frac{m}{2} \left(g - \frac{R\omega^2 h_c}{l_c} \right) = \frac{m}{2} \left(g - \frac{v^2 h_c}{R l_c} \right) \tag{7.121}$$

and

$$N_B = \frac{m}{2} \left(g + \frac{R\omega^2 h_c}{l_c} \right) = \frac{m}{2} \left(g + \frac{v^2 h_c}{R l_c} \right) , \tag{7.122}$$

where we have employed $v = R\omega$.

From Eqs. (7.121) and (7.122), we observe that as v increases, N_A will vanish as the speed of the car reaches a critical v_c. At that instant, wheel A will lose contact with the ground. By setting $N_A = 0$, we obtain the critical v_c:

$$v_c = \sqrt{\frac{gRl_c}{h_c}} . \tag{7.123}$$

This result shows that in order to increase the critical speed v_c which would make the car more difficult to lose contact with the road, we need to increase the traction width l_c and reduce the height h_c of the car. Such a configuration is typically employed in the design of sport cars. Moreover, Eq. (7.123) shows that when a car turns in a tight corner, the smaller R of such a corner leads to a smaller v_c, making turning relatively more unstable.

What if the speed of the car goes beyond v_c? At which point, we know that $N_A = 0$ and $F_{sA} = 0$. Equations (7.117) and (7.118) then implies that $N_B = Mg$ and $F_{sB} = MR\omega^2$ respectively. The magnitude of the anti-clockwise torque is then Mgl_c. On the other hand, re-expressing $F_{sB} = Mv^2/R$ implies a total clockwise torque of Mh_cv^2/R. Because $v > v_c$, we have $v^2 > gRl_c/h_c$, which means that $Mh_cv^2/R > Mgl_c$. This shows that the clockwise torque is greater than the anti-clockwise torque, with the presence of a net torque and an angular acceleration in the clockwise direction. Thus, when the speed of the car goes beyond v_c, we expect the car to roll over in the clockwise direction. There is no circumstances under the given conditions that the car can lean and turn like the rider-motorcycle system.

7.3.7 *Final Examples on Rotational Dynamics*

In this section, we shall illustrate the application of the theory developed in this chapter to three different versions of rotating sticks.

7.3.7.1 *A Falling Stick without Pivot*

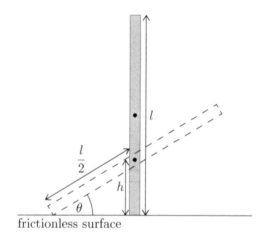

frictionless surface

Fig. 7.24

In the first version, we have a vertical stick of length l at rest precariously on a frictionless ground as shown in Fig. 7.24. A very slight perturbation then causes the stick to move. Because there are only the weight mg of the stick and the normal force N from the ground that act on the stick, and since these two forces are purely vertical, the COM of the stick only moves vertically downnward without any horizontal motion while the stick slides on the ground and rotates about its COM. Let us now determine the final angular speed of the stick just before it touches the ground. To do that, we first obtain a relation between the vertical distance of the COM from the ground h and the angle θ between the orientation of the stick to the horizontal (see Fig. 7.24):

$$h = \frac{l}{2} \sin \theta$$

Taking the differential of h and θ from this equation, we obtain

$$dh = \frac{l}{2} \cos \theta d\theta \,,$$

which allows us to deduce the COM velocity of the stick as follows:

$$\vec{v}_{CM} = \frac{dh}{dt} = \frac{l\omega}{2} \cos \theta \,. \tag{7.124}$$

Equation (7.127) informs us that at the moment when the stick touches the ground, i.e., when $\theta = 0$,

$$\vec{v}_{CM} = l\omega/2 \,. \tag{7.125}$$

In order to determine the angular speed ω of the stick just before it hits the ground, we employ the approach of conservation of energy. Initially, the stick only possesses $mgl/2$ amount of gravitational potential energy. By the time it is about to hit the ground, this potential energy has been converted entirely to kinetic energy in the form of both translation and rotation as described by the following equation:

$$mg \left(\frac{l}{2} \right) = \frac{1}{2} m \vec{v}_{CM}^2 + \frac{1}{2} I_{CM} \omega^2 \,,$$

$$mg \left(\frac{l}{2} \right) = \frac{1}{2} m \left(\frac{l^2 \omega^2}{4} \right) + \frac{1}{2} \left(\frac{1}{12} m l^2 \right) \omega^2 \,. \tag{7.126}$$

Solving the above equation for ω, we arrive at

$$\omega = \sqrt{\frac{3g}{l}} \,. \tag{7.127}$$

7.3.7.2 *A Falling Stick with a Pivot*

The falling stick in this version modifies that of the last section with the stick being fixed to a pivot at the point where it touches the ground. In other words, the stick in this example is constrained by the pivot as it falls. Other than the mg of the stick and a normal force from the pivot, there is

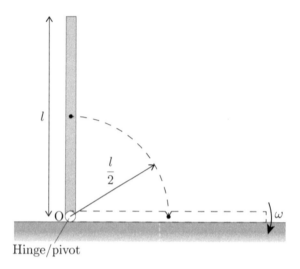

Fig. 7.25

also an additional horizontal force from the pivot on the stick. Thus, the COM of the stick does not just fall vertically like in section 7.3.7.1, it also moves horizontally. In fact, the COM rotates in a circular motion about the pivot point O with a radius of $l/2$ as it falls from rest from its starting vertical orientation till it reaches a horizontal bearing (see Fig. 7.25) just before hitting the ground. As in section 7.3.7.1, we are to determine the angular speed of the stick at the instant before it hits the ground. Like before, we adopt the approach of conservation of energy, with the initial gravitational potential energy of $mgl/2$ being entirely converted to rotational kinetic energy about point O, which is described by the following equations:

$$mg\left(\frac{l}{2}\right) = \frac{1}{2}I_O\omega^2,$$

$$mg\left(\frac{l}{2}\right) = \frac{1}{2}\left(\frac{1}{3}ml^2\right)\omega^2. \tag{7.128}$$

Solving the last equation for ω, we obtain

$$\omega = \sqrt{\frac{3g}{l}}. \tag{7.129}$$

This angular speed is the same as Eq. (7.127) in the last section, which is expected, since the constraint forces from the pivot do not do work against the stick during its rotation (the torque of the constrained forces about point O is zero).

7.3.7.3 *A Stick that is Pivoted and Shot by a Bullet*

In our final version, we have a uniform stick of mass m_s and length d being pivoted at the center. The stick is initially at rest. A bullet of mass m_b is shot through the stick at a point halfway between the pivot and the bottom end of the stick (refer to Fig. 7.26). The initial speed of the bullet is v_i, and its final speed is v_f. Our goal is to determine the angular speed ω_f of the stick just after the collision.

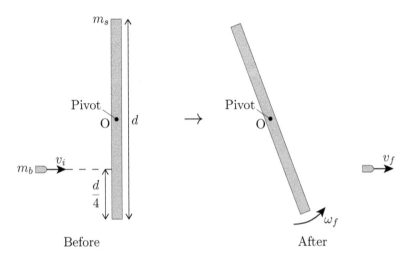

Fig. 7.26

Since there is net external force at the pivot, the principle of conservation of linear momentum is not applicable for this problem. However, if we were to take moments about point O which is located at the stick's COM, there is an absence of external torque. This implies the applicability of conservation of angular momentum about point O:

$$L_i = L_f . \tag{7.130}$$

Note that L_i is the initial angular momentum of the system before collision, and L_f is the final angular momentum of the system after collision. Both angular momentum are taken about the point O. Since only the bullet is moving before the collision, the initial angular momentum is solely contributed by the bullet:

$$L_i = m_b v_i \left(\frac{d}{4}\right) . \tag{7.131}$$

After the collision, the speed of the bullet has reduced and in concomitant, the stick now rotates with an agular speed of ω_f, giving rise to the following angular momentum of the system:

$$L_f = m_b v_f \left(\frac{d}{4}\right) + I_{CM}\omega_f . \tag{7.132}$$

Substituting Eqs. (7.131) and (7.132) into Eq. (7.130), we have

$$m_b v_f \left(\frac{d}{4}\right) + I_{CM}\omega_f = L_i = m_b v_i \left(\frac{d}{4}\right) .$$

Solving for ω_f and noting that $I_{CM} = m_s d^2/12$ (where we have assumed that the small hole in the stick due to the entry of the bullet has negligible effect on the stick's moment of inertia), we proceed to evaluate

$$\omega_f = m_b (v_i - v_f) \left(\frac{d}{4I_{CM}}\right) ,$$

$$= m_b (v_i - v_f) \left(\frac{d}{4}\right) \left(\frac{12}{m_s d^2}\right) ,$$

which finally leads to

$$\omega_f = \frac{3m_b (v_i - v_f)}{m_s d} . \tag{7.133}$$

Appendix

A.1 Fictitious Forces in a Rotating Reference Frame of Constant Angular Velocity

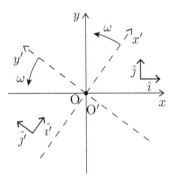

Fig. A.1

Let us consider two Cartesian x-y plane in this section. The first plane S is an inertial reference frame and is shown with its x-y axes fixed in space in Fig. A.1. The second plane S' is a non-inertial reference frame that rotates counter-clockwise about its origin O' (which coincides with the origin of O of S) with a constant angular velocity ω. The angular velocity of S' is in fact a vector $\omega \hat{k}$, with $\hat{k}$ being a unit vector pointing out of the page of the book. Then, the spatial position of a particle can be referenced by its position vector $\vec{r}$ in the plane as follows:

$$\vec{r} = \tilde{r}_x \hat{i} + \tilde{r}_y \hat{j} = \tilde{r}_{x'} \hat{i}' + \tilde{r}_{y'} \hat{j}', \qquad (A.1)$$

where the unprimed representation depicts the position of the particle with respect to the inertial reference frame S, while the primed representation with that of the non-inertial reference frame S'. Note that

$$\tilde{r}' = R\tilde{r}, \tag{A.2}$$

with

$$\tilde{r} = \begin{pmatrix} \tilde{r}_x \\ \tilde{r}_y \end{pmatrix}, \tag{A.3}$$

$$\tilde{r}' = \begin{pmatrix} \tilde{r}_{x'} \\ \tilde{r}_{y'} \end{pmatrix}, \tag{A.4}$$

and

$$R = \begin{pmatrix} \cos\omega t & \sin\omega t \\ -\sin\omega t & \cos\omega t \end{pmatrix}. \tag{A.5}$$

Similarly, the velocity $\vec{v}$ and acceleration $\vec{a}$ of the particle will be depicted by $\tilde{v}$ and $\tilde{a}$ with respect to S, and $\tilde{v}'_I$ and $\tilde{a}'_I$ with respect to S'. In the same sense as Eq. (A.2), we would expect the components of velocity and acceleration in the S frame to depend on those in the S' frame via the following transformation:

$$\tilde{v}'_I = R\tilde{v}, \tag{A.6}$$

and

$$\tilde{a}'_I = R\tilde{a}. \tag{A.7}$$

Notice that we have inserted a subscript I to both the velocity and acceleration terms of the reference frame S'. This is to emphasize that these terms are referring to the velocity and acceleration vectors captured by the inertial reference frame S, but using the representation of S' to describe them here[7].

Before we proceed, let us yield a few mathematical properties of the rotation matrix R that will be useful for our purpose. First, R is an orthogonal matrix that satisfies the following properties:

$$RR^T = I, \tag{A.8}$$

where the superscript T corresponds to the matrix transpose operation and I is the identity matrix. Because R is time-dependent, we can differentiate the above equation with respect to time to arrive at the following relation:

$$\frac{dR}{dt}R^T + R\left(\frac{dR}{dt}\right)^T = 0, \tag{A.9}$$

---

[7]This means literally that $\vec{v} = \tilde{v}_x\hat{\imath} + \tilde{v}_y\hat{\jmath} = \tilde{v}^I_{x'}\hat{\imath}' + \tilde{v}^I_{y'}\hat{\jmath}'$ and $\vec{a} = \tilde{a}_x\hat{\imath} + \tilde{a}_y\hat{\jmath} = \tilde{a}^I_{x'}\hat{\imath}' + \tilde{a}^I_{y'}\hat{\jmath}'$.

where we have used the fact that $d\left(R^T\right)/dt = (dR/dt)^T$. Let us define the matrix S as

$$\frac{dR}{dt}R^T = S \tag{A.10}$$

and substitute it into Eq. (A.9) to obtain

$$S + S^T = 0\,,$$

which implies that S is a skew-symmetric matrix. In fact, it is easy to evaluate the form of S from its definition, which is

$$S = \begin{pmatrix} 0 & \omega \\ -\omega & 0 \end{pmatrix}\,. \tag{A.11}$$

Furthermore, S is found to satisfy the following property:

$$S^2 = SS = -\omega^2 I \tag{A.12}$$

Now, let us use the orthogonal properties of R as given by Eq. (A.8) on Eq. (A.10) to obtain the following form:

$$\frac{dR}{dt} = SR\,. \tag{A.13}$$

We are now ready to determine the effect of the non-inertial reference frame on the velocity and acceleration of the particle. For the velocity, let us differentiate both sides of Eq. (A.2) with respect to time:

$$\frac{d\tilde{r}'}{dt} = R\frac{d\tilde{r}}{dt} + \frac{dR}{dt}\tilde{r}\,,$$

$$\tilde{v}' = R\tilde{v} + \frac{dR}{dt}\tilde{r}\,,$$

$$\tilde{v}' = R\tilde{v} + SR\tilde{r}\,, \tag{A.14}$$

where we have used Eq. (A.13) in the last line. By comparing this equation with Eq. (A.6), we observe that the term $SR\tilde{r}$ is a consequence of the non-inertial reference frame. Since Eq. (A.14) can be expressed explicitly as

$$\tilde{v}' = \begin{pmatrix} \tilde{v}^I_{x'} + \omega\tilde{r}_{y'} \\ \tilde{v}^I_{y'} - \omega\tilde{r}_{x'} \end{pmatrix}\,, \tag{A.15}$$

after using Eqs. (A.2), (A.6) and (A.11), we can write

$$\begin{aligned}
\vec{v}' &= \tilde{v}_{x'}\hat{i}' + \tilde{v}_{y'}\hat{j}'\,, \\
&= \left(\tilde{v}^I_{x'} + \omega\tilde{r}_{y'}\right)\hat{i}' + \left(\tilde{v}^I_{y'} - \omega\tilde{r}_{x'}\right)\hat{j}'\,, \\
&= \left(\tilde{v}^I_{x'}\hat{i}' + \tilde{v}^I_{y'}\hat{j}'\right) + \left(\omega\tilde{r}_{y'}\hat{i}' - \omega\tilde{r}_{x'}\hat{j}'\right)\,, \\
&= \vec{v} + \left(\omega\tilde{r}_{y'}\hat{i}' - \omega\tilde{r}_{x'}\hat{j}'\right)\,.
\end{aligned} \tag{A.16}$$

Because

$$\vec{r} \times \vec{\omega} = \begin{vmatrix} \hat{\imath}' & \hat{\jmath}' & \hat{k}' \\ \tilde{r}_{x'} & \tilde{r}_{y'} & 0 \\ 0 & 0 & \omega \end{vmatrix},$$

$$= \omega \tilde{r}_{y'} \hat{\imath}' - \omega \tilde{r}_{x'} \hat{\jmath}', \tag{A.17}$$

we can write Eq. (A.16) as

$$\vec{v}' = \vec{v} + \vec{r} \times \vec{\omega}, \tag{A.18}$$

where the term $\vec{r} \times \vec{\omega}$ results from the effect of the non-inertial reference frame.

For the acceleration, let us differentiate both sides of Eq. (A.14) with respect to time:

$$\frac{d\tilde{v}'}{dt} = R\frac{d\tilde{v}}{dt} + \frac{dR}{dt}\tilde{v} + SR\frac{d\tilde{r}}{dt} + S\frac{dR}{dt}\tilde{r},$$

$$\tilde{a}' = R\tilde{a} + \frac{dR}{dt}\tilde{v} + SR\tilde{v} + S\frac{dR}{dt}\tilde{r},$$

$$\tilde{a}' = R\tilde{a} + SR\tilde{v} + SR\tilde{v} + S^2R\tilde{r},$$

$$\tilde{a}' = R\tilde{a} + 2SR\tilde{v} - \omega^2 R\tilde{r}, \tag{A.19}$$

where we have employed Eqs. (A.12) and (A.13). In order to put the right-hand side of the above equation with respect to the S' frame, we first multiply both sides of Eq. (A.14) with S:

$$S\tilde{v}' = SR\tilde{v} + S^2R\tilde{r},$$

$$S\tilde{v}' = SR\tilde{v} - \omega^2 R\tilde{r},$$

where we have again employed Eq. (A.12) in the last line. This equation implies

$$SR\tilde{v} = S\tilde{v}' + \omega^2 R\tilde{r}.$$

Substituting this equation into Eq. (A.19) we obtain

$$\tilde{a}' = R\tilde{a} + 2\left(S\tilde{v}' + \omega^2 R\tilde{r}\right) - \omega^2 R\tilde{r},$$

$$\tilde{a}' = R\tilde{a} + 2S\tilde{v}' + \omega^2 R\tilde{r}. \tag{A.20}$$

By comparing Eq. (A.20) with Eq. (A.7), we observe that the terms $2S\tilde{v}' + \omega^2 R\tilde{r}$ is a consequence of the non-inertial reference frame. We can express Eq. (A.20) as follows:

$$\tilde{a}' = \begin{pmatrix} \tilde{a}^I_{x'} + 2\omega\tilde{v}_{y'} + \omega^2\tilde{r}_{x'} \\ \tilde{a}^I_{y'} - 2\omega\tilde{v}_{x'} + \omega^2\tilde{r}_{y'} \end{pmatrix}, \tag{A.21}$$

which allows us to write

$$\vec{a}' = \tilde{a}_{x'}\hat{i}' + \tilde{a}_{y'}\hat{j}',$$
$$= \left(\tilde{a}_{x'}^I + 2\omega\tilde{v}_{y'} + \omega^2\tilde{r}_{x'}\right)\hat{i}' + \left(\tilde{a}_{y'}^I - 2\omega\tilde{v}_{x'} + \omega^2\tilde{r}_{y'}\right)\hat{j}',$$
$$= \left(\tilde{a}_{x'}^I\hat{i}' + \tilde{a}_{y'}^I\hat{j}'\right) + \left(2\omega\tilde{v}_{y'}\hat{i}' - 2\omega\tilde{v}_{x'}\hat{j}'\right) + \omega^2\left(\tilde{r}_{x'}\hat{i}' + \tilde{r}_{y'}\hat{j}'\right),$$
$$= \vec{a} + 2\left(\omega\tilde{v}_{y'}\hat{i}' - \omega\tilde{v}_{x'}\hat{j}'\right) + \omega^2\vec{r}. \tag{A.22}$$

Since

$$\vec{v}' \times \vec{\omega} = \begin{vmatrix} \hat{i}' & \hat{j}' & \hat{k}' \\ \tilde{v}_{x'} & \tilde{v}_{y'} & 0 \\ 0 & 0 & \omega \end{vmatrix},$$
$$= \omega\tilde{v}_{y'}\hat{i}' - \omega\tilde{v}_{x'}\hat{j}', \tag{A.23}$$

we have

$$\vec{a}' = \vec{a} + 2\vec{v}' \times \vec{\omega} + \omega^2\vec{r}. \tag{A.24}$$

Note that the last two terms result from the non-inertial reference frame with $2\vec{v}' \times \vec{\omega}$ being the Coriolis acceleration and $\omega^2\vec{r}$ being the centrifugal acceleration. By multiplying both sides of Eq. (A.24) with m, we obtain

$$m\vec{a}' = m\vec{a} + 2m\vec{v}' \times \vec{\omega} + m\omega^2\vec{r}. \tag{A.25}$$

Because $\vec{a}$ is with respect to the inertial reference frame, $m\vec{a} = \vec{F}_{real}$ based on Newton's second law of motion. Therefore, with respect to a non-inertial reference frame that is rotating with a constant angular velocity $\vec{\omega}$, Newton's second law of motion is written as follows:

$$\vec{F}_{real} + 2m\vec{v}' \times \vec{\omega} + m\omega^2\vec{r} = m\vec{a}', \tag{A.26}$$

with $2m\vec{v}' \times \vec{\omega} + m\omega^2\vec{r}$ being the fictitious forces. Note that $2m\vec{v}' \times \vec{\omega}$ is the Coriolis force, while $m\omega^2\vec{r}$ is the centrifugal force.

Index

acceleration, 17
  average, 17
  average angular, 106
  centripetal, 28
  instantaneous, 17
  instantaneous angular, 106
  radial, 28, 108
  relative, 30
  tangential, 28, 108
action and reaction pair, 42
angular displacement, 105
angular momentum, 109
angular velocity, 27
appendix, 153

Cartesian coordinate system, 7
  transformation between the polar
    coordinates and the
    Cartesian coordinates, 9
center of mass
  (COM) frame, 89, 117
  acceleration, 86
  and Newton's second law, 86
  for continuous mass distribution,
    85
  for discrete masses, 84
  relation to conservation of linear
    momentum, 85
  velocity, 86
centrifugal force, 52
centripetal acceleration, 28
centripetal force, 48

coefficient of restitution, 84
collisions
  elastic, 82
  inelastic, 83
  perfectly inelastic, 83
conservation laws:
  of angular momentum, 110
  of energy, 61
  of linear momentum, 80
  of mechanical energy, 69
conservative force, 71
coordinate system
  Cartesian, 7
  polar, 9
  transformation between the polar
    coordinates and the
    Cartesian coordinates, 9
Coriolis force, 52

displacement, 13
  angular, 105
  position vector, 14
  relative, 30
distance, 15
dot product, 4

elastic collision, 82
energy
  conservation, 61
  conservation of mechanical energy,
    69
  kinetic, 69

potential, 69
  rotational kinetic, 115
environment, 61
external force, 80

fictitious force, 52
first law of thermodynamics, 61
force, 39
  centrifugal, 52
  centripetal, 48
  conservative, 71
  Coriolis, 52
  external, 80
  fictitious (pseudo-force), 51, 52
  impulsive, 82
  internal, 80
  nonconservative, 72
  resistive, 44, 47
friction, 44
  kinetic, 45
  rolling, 129
  static, 44

Galilean coordinate transformation,
  37
Galilean velocity transformation, 38

Hooke's law, 65

impulse
  definition, 81
  Impulse-Momentum Theorem, 81
impulsive approximation, 82, 87
impulsive force, 82
inelastic collision, 83
inertia, 40
  law of, 40
  moment of, 113
internal force, 80
irreversibility, 47

kinematics, 13
  kinematic equations, 18
  rotational kinematic equations, 107
kinetic energy, 69
  rotational, 115

kinetic friction, 45

law of inertia, 40
linear momentum, 79

mass, 40
moment arm, 114
moment of inertia, 113
momentum
  angular, 109
  conservation of angular
    momentum, 110
  conservation of linear momentum,
    80
  linear, 79

Newton's first law, 40
Newton's second law
  expressed in terms of linear
    momentum, 94
  rotational analog of, 113
  translational case, 41
Newton's third law
  and conservation of linear
    momentum, 80
  statement, 42
  third-law (action and reaction)
    pair, 42
nonconservative force, 72

parallel axis theorem, 121
parallelogram law, 3
perfectly inelastic collision, 83
polar coordinate system, 9
  transformation between the polar
    coordinates and the
    Cartesian coordinates, 9
potential energy, 69
principle of relativity, 37
projectile motion, 19, 31
pseudo-force, 52
pseudowork, 76
  definition, 77
  examples:
      done by kinetic friction on a
      rotating sphere, 140

pure rolling motion of an
  accelerating car wheel, 134
  rolling down an inclined
    plane, 129
  rolling up an inclined plane,
    133

quasi-static, 66, 67

radial acceleration, 28, 108
reference frame, 29
  inertial, 37
  non-inertial, 51
relative acceleration, 30
relative displacement, 30
relative velocity, 30
resistive force, 44, 47
right-hand rule, 6, 106
rigid body, 103
rolling, 126
  with slipping, 136
  without slipping, 126
rolling friction, 129
rotational kinetic energy, 115

scalar, 1
scalar product, 4
static friction, 44
system, 61
  isolated, 68

tangential acceleration, 28, 108
time
  time instant, 13
  time interval, 13
torque (definition), 109

vector, 1
  addition, 2
    associative law, 3
    commutative law, 3
  basis, 7
  cross product, 5
  dot (scalar) product, 4
    commutative law, 5
    distributive law, 5
  multiplication, 4
  multiplied by a scalar, 4
  negative, 3
  subtraction, 4
  zero, 3
velocity, 15
  angular, 27
  average, 15
  average angular, 105
  instantaneous, 16
  instantaneous angular, 106
  relative, 30

work done
  and elastic potential energy, 68
  by constant force, 62
  by variable force, 64
  elastic spring, 66
Work-Energy Theorem, 69
  for rotational motion, 117

CPSIA information can be obtained
at www.ICGtesting.com
Printed in the USA
FSHW010502180820
73041FS